Structural Engineering Documents

Franz Knoll and Thomas Vogel

Design for Robustness

International Association for Bridge and Structural Engineering IABSE
Association Internationale des Ponts et Charpentes AIPC
Internationale Vereinigung für Brückenbau und Hochbau IVBH

ISBN 978-3-85748-120-8
DOI: https://doi.org/10.2749/sed011

Printed in Switzerland

Publisher:
IABSE-AIPC-IVBH
ETH Zurich
CH-8093 Zurich, Switzerland

Phone: Int. + 41-44-633 2647
Fax: Int. + 41-44-633 1241
E-mail: secretariat@iabse.org
Web: www.iabse.org

Preface

Robustness is a property, the description of which varies so much with context that it is difficult to put order into its manifold aspects, relationships and ramifications, let alone to pronounce a consistent and general theory.

This text is an attempt to provide at least a practical review of the important elements of robustness in the context of structural systems, and to collect ideas or ways and means to deal with some typical circumstances in terms of structural design, in order to enhance survival, or to mitigate the consequences of unforeseen events to structural systems.

The text is divided into two parts:

- A review of the elements of robustness and strategies for its establishment by design (Chapters 1–8). Hopefully, this part will be found to be sufficiently short and concise for the reader not to become overly and prematurely bored.
- A review of specific scenarios intended to illustrate some typical or notorious situations where robustness must be established beyond the schoolbook structural design procedures (Chapters 9 and 10).

Writing a book is a lonely business. It's all the more important that other people have an attentive look to the outcome before it is printed. The authors wish to express their gratitude to the IABSE Structural Engineering Documents Editorial Board with its chairmen Geoff Taplin and Mikael W. Braestrup and their assigned reviewer Loring A. Wyllie, Jr who spent their valuable time in reading the manuscript and helped with their feedback to improve the present document. Finally, special thanks go to IABSE and its headquarter who made it possible to disseminate our experiences and ideas to such a distinguished audience.

Franz Knoll
Nicolet Chartrand Knoll, Montreal, QC, Canada

Thomas Vogel
Institute of Structural Engineering, ETH Zurich, Switzerland

Table of Contents

Chapter

1 Introduction

1.1 What is robustness?

Robustness is the property of systems that enables them to survive unforeseen or unusual circumstances.

The design of a system, be it a natural or an artificial one, is typically oriented towards normal use, i.e. circumstances which must or can be anticipated to exist during the intended working life of the system. Limiting the design to this may however leave it vulnerable to the effects of events that were not included in the set of anticipated circumstances. These effects can be of very diverse character and may be related to the features that were anticipated in the design but for an unanticipated intensity, or they may be of a description altogether foreign to the design premises.

The first case can be exemplified by a structure, which, although it was designed to resist a set of physical conditions (e.g. climatic or seismic effects), succumbed to some of these because they turned out to be of greater magnitude than foreseen.

An example for the second case could be seen in the fate of natural systems, for instance certain species that are perishing because of exposure to anthropogenic adversity in the form of destruction of habitat or such. Nature will, of course, always survive and mostly this will happen through expansion of the system: Where one species as a system is lost, others will take its place as part of a larger system—nature can expand *the system* up to a planetary scope and beyond—"Nature will survive humanity". The question for us is of course whether humanity will survive, and how, with humanity's power to expand itself and the systems it creates being limited.

In order to say something rational and consistent on the property of robustness, some basic concepts must be described and clarified as far as possible—although a strict definition in the sense of a reduction onto other, well-known concepts may just be out of reach. The two concepts foremost in need of clarification relate to two terms used in the brief introduction above: *System* and *Survival*.

1.2 System

In the case of *system*, it is easy to see that this term can mean anything in a very wide range of variations. From natural ecosystems of different scope to the digestive apparatus of a living individual, and from the political system of a state to a tool such as a computer or a kitchen appliance to a mental construct such as philosophy, or the instructions for the use of that kitchen appliance. In view of this diversity, it would appear that limiting the scope of application would provide the only hope for finding a manageable description of *system*. So here goes: Since the goal of this report is to find ways to make structures more robust, the concept of system shall be limited to what can reasonably be called a structural system. The term still includes a considerable variety of things, from the cast-in-place concrete frame of a building to a method of supporting formwork for the same, to an individual structural element such as a bridge girder or welded steel connections. All of these constitute subsystems to an overall structural system and their robustness will be instrumental for the survival of the latter.

Unlike in nature, all structural systems are limited in time, space and purpose. While in natural systems the passing of an individual is of no great importance, this is not the case for human artefacts, which are created and purchased for a distinct function. If the roof of my house collapses under the weight of the 50-year snow during a blizzard, its replacement is not an acceptable recourse since it was built to shelter me from snow and cold and wind, i.e. for a distinct and limited purpose.

1.3 Survival

The other concept essential for the discussion of robustness is *survival*. Survival is not an absolute and its description may vary, depending on the context. Normally it means *survival of function*, i.e. through its robustness, the structural system must continue to provide the function for which it was created, modified or preserved, and it must do this *whatever happens*, i.e. independent of circumstances. These circumstances may include limited damage to the structural system, perhaps a reduction or interruption of the full function limited in time, but essentially the function must be maintained through the intended working life of the structural system. A building, which comes through an earthquake with some cracks, broken glass or the like but can be repaired in reasonable time and at acceptable cost, has survived even if some of its occupants must be evacuated or inconvenienced for a while. Not so for another building, which, although it is still standing up, must be condemned and demolished because it would take too long, be too costly or too dangerous to repair it.

For the discussion ahead, the term *structural* then needs to be substantiated also. Structural functions normally include: Resistance to load effects or chemical attack, shelter from climatic phenomena, containment of substances, and sometimes more specialized purposes such as providing visual aspects, fortification, security, shade, etc.

1.4 Robustness in structural codes

Although some building codes require that structures should be robust, only the newest ones define robustness in a prominent place. The Eurocodes for instance that should replace the National building codes of all countries of the European Community and some others such as Norway, Switzerland, Iceland, Cyprus require robustness in their Basis of Design only implicitly ([3] Clause 2.1 Basic requirements):

"[3] A structure shall be designed and executed in such a way that it will not be damaged by events such as:

- explosion,
- impact, and
- the consequences of human errors.

to an extent disproportional to the original cause.

[4] Potential damage shall be avoided or limited by appropriate choice of one or more of the following:

- avoiding, eliminating or reducing hazards to which a structure can be subjected;
- selecting a structural form which has a low sensitivity to the hazards considered;
- selecting a structural form and design that can survive adequately the accidental removal of an individual member or a limited part of the structure, or the occurrence of acceptable localised damage;
- avoiding as far as possible structural systems that can collapse without warning;
- tying the structural members together."

In Eurocode 1–7 on accidental actions, finally, robustness is defined as: "the ability of a structure to withstand events like fire, explosions, impact or the consequences of human error, without being damaged to an extent disproportionate to the original cause" ([4], Clause 1.5.14).

It is of course easy for a committee to write such sweeping requirements into the code. It leaves the engineer in a rather uncomfortable situation, however, if help is not provided along with the demand. No code is presently doing this in any useful way, leaving the engineers to themselves with the task. This little book is intended to provide some of that help.

Chapter

2

The Foreseeable Unforeseen

2.1 Ordinary structural design

Today, the design of structural systems is most often based on *mathematical models*, which are substituted for the future *real structure*, for purposes of analysis. Modelled are, besides the physical description of the structure (geometry, topology, stiffness, mass, weight, etc.), the circumstances to which the structure is thought to become exposed during its lifetime (loading effects, chemical deterioration, abrasion, etc.). Sometimes, physical models such as scaled down reproductions, full-scale prototypes or the final structure itself are used to justify its design, through proof loading or other representative testing.

The fact that the proof of aptness of the structural system must be brought about using substitutes of some form and can only rarely be demonstrated on the final product[1],has been recognized and rationalized by the *règles de l'art* (i.e. *the art of engineering*), and expressed in building codes and handbooks in the form of modelling rules and stipulated minimum safety margins. Safety margins are intended to compensate for differences which are anticipated to exist between the idealized models used for the analysis, and the physical equivalent, the real structure and its circumstances. Where the maximum snow load expected to occur in 50 years is quoted to 2.5 kPa in the building code, a satisfactory design must provide a structure which will resist a higher load (e.g. 1.5 times). At the same time, the modelled resistance of the structure is put into question and somewhat lower resistance values are presumed in the design, recognizing that the creation of the real structure may leave something to be desired and may result in a somewhat less resistant system than specified on paper.

These safety margins have been developed, mainly through a vast amount of experience with real structures but also through much research in laboratories, and their values are usually established as a consensus in committees appointed for the purpose. They are periodically adjusted, reflecting new information concerning the aptness of their values and also their format. However, for the great majority of applications, no substantial adjustments (greater than, e.g. 10%) have been introduced recently, reflecting a general impression that all is as it should be.

[1]Often structural systems are, unlike industrial products, "one of a kind", so that no prototype or serial testing is possible, especially when it is destructive.

2.2 Shortcomings

This may be quite so as far as the values and the format of the safety margins go, but on the other hand, accidents still happen, not only to the chagrin of the people who are directly involved but also to the detriment of the economy which will have to come up with the goods to remedy the loss.

In the forensic investigation following accidents, it is often recognized that the circumstances of the accident have little to do with the numerical safety margins which were used in the design of the failed structural system, but must be attributed to something which was not anticipated altogether at the time and in the context of design and analysis. In other words, a mere amplification of safety margins might not have prevented the accident.

Since most of the time the external circumstances (outside the structural system) causing an accident cannot be eliminated — from extreme natural events to accidental or hostile action of some human individuals or groups — the goal is to find ways to endow the structural system with the property of robustness so that it will survive these unforeseen events. Let us not forget also that the structural system itself may contain flaws, which can cause accidents, external circumstances being innocently within their normal, anticipated range. Most often, human error is cited to be at the origin of such flaws — human error has always been with us and will not stop doing so; it must be assumed. So again: Ways must be found for structural systems to survive the unforeseen, even if this may be a flaw in the system itself.

2.3 Answers from our ancestors

Historically, structural design had to be done without the benefit of mathematical models and analytical calculations, let alone computers. And classically the builders, "mere" practitioners as it were, have arrived at an impressive menu of safe structural systems, most of them equipped with a sufficient degree of robustness to make them endure for a useful length of time. In contrast, numerous cases are known where structures, although analysed with great care and designed to the most advanced theories of their time, have failed miserably due to circumstances that were not reflected in the theory or in the modelling.

What is it then that makes some structures endure which were created without the help of any rational theory, whereas others fail in spite of the great effort which went into the creation and application of the theory that was used for the design — although of course we should not forget that the failures of earlier times are no longer here to show. . .?

The answer to this question is varied and manifold:

"One bolt is no bolt" our teacher in steel construction used to say. So even if the theory says that the strength of a single bolt is sufficient to transfer the anticipated forces from one element of the structure to the next, let us provide at least two of them because the one single bolt could just be the bad one of the batch, breaking unannounced. The consequences of that rupture are not comparable to the cost of an additional bolt, which by consequence becomes a worthwhile investment. The rule must be traced back to the century-old construction practice where every builder knew that it paid off to provide some extra strength, "just in case", and not to "put all eggs in the same basket".

History has tales of instances where this rule was ignored, followed by catastrophic failures (slum construction in ancient Rome, Beauvais cathedral, modern concrete bridges built in the 1960s and 1970s). These were designs which were deliberately minimized or "optimized", to the *bare bones* or the *limits* of what one could still pretend to be credible, only to fail miserably because a little something occurred which took the role of the proverbial straw breaking the camel's back.

We now have the tools to model all of this and to assess the physical consequences (forces, deformations, weakening) and probability of the same, as well as to analyse the response of the structure we are designing — if only everything is being considered correctly in the analysis and nothing important is left out.

Some structural systems are intentionally conceived to suffer damage in the anticipated circumstances but to continue fulfilling their function. As an example for this, the guardrails along auto routes can be cited which will not resist being bent and twisted in a forceful collision but must prevent the vehicle from leaving the roadway.

Some structural systems may even change function as a consequence of certain circumstances: A car in an accident changes from a means of transportation to one of protection, sheltering its occupants from physical harm. Likewise, buildings will, in an extreme earthquake, be allowed to suffer some damage, even impairing their primary function as housing or workplace if only they continue providing safe exit or serving as a shelter from falling debris, fire, etc.

Chapter

3 Survival through Robustness

For a structural system to survive unforeseen events or circumstances, with its intended function intact, it must possess sufficient reserve capacity to stand up to conditions during and after the event. A robust structural system has therefore:

$$\text{Residual capacity} \geq \text{Residual demand} \tag{3.1}$$

Most often, *capacity* will relate to resistance to forces (i.e. strength), but it may also mean deformability, ductility, stability, weight, mass or stiffness as any of these properties may be critical, depending on the context. The term *residual* or *after the event* may not always be temporal in the literary sense but, as in the case of a hidden flaw or weakness, may rather mean: In the situation where the event (the flaw taking effect) has taken place.

A structural system may change its character during or due to *the event*, in particular concerning the description of *load paths*. This term can be defined as the integral of all elements of the system affected by internal and external forces. It is described by stresses, internal forces, reactions, etc., which occur in those elements and can be traced by calculating or measuring those quantities from the point of application of the load to the boundaries of the system. A simple example is provided by the chain with a weight hanging from it: The load path goes from the hook through all the chain links to the point of suspension (and beyond if elements beyond are to be included in the system).

The example of the guardrail along the roadway is quite illustrative for the change in load path: When the impact of a collision is light, the rail will resist it in bending, without being deformed much. Heavier impacts will be resisted in membrane fashion, the guardrail being bent out of position and acting now mainly in a hammock-like manner. This second load path represents a *second line of defence*, one of the principal strategies for robustness.

As indicated above, the structural system may also change its function as the result of the event, as in the example of the car in an accident, or a building after an extreme earthquake. This is not always the case, however, and structures may have to be designed to survive with their primary function intact, or with a similar one assigned to them after the event. Important examples for this are – aside from military facilities – hospitals, schools or similar spaces intended to serve as shelters for people who were displaced by the event. Likewise, roadways, bridges and airports needed for access to distressed areas, must serve with their original function preserved following the event. If they do not, considerable loss of life is often the consequence.

Another class of structures that must come through an event with their primary function unimpaired includes all facilities containing dangerous substances, the escape or ignition of which would be catastrophic. This concerns nuclear power installations, tanks and containers of flammable or poisonous liquids or gases, explosives or dams the failure of which will cause flooding. Modern building codes and design manuals recognize this by stipulating more restrictive design parameters for a selection of cases.

Robustness of a structural system is therefore very much a matter of context or of the scenario of the event, and the design and analysis of robustness must be adapted to the particular situation. In cases where information on the event is scarce – such as with the hidden flaw – robustness will have to follow a *shot gun approach* and be provided in a general sense, considering the possible weakness of any element of the structural system.

A number of notorious cases persist where today's technology has not been able (or society willing) to come up with ways to install sufficient structural robustness. These are still causing accidents of catastrophic magnitude such as train and airplane crashes, shipwrecks or the spill of poisonous materials following derailment, fire, explosion or impact.

Catastrophic earthquakes are beginning to have their effects mitigated, especially in the wealthy countries, but they are still taking full effect in some regions of the planet where society does neither have the resources nor the technical knowledge to take remedial action and replace vulnerable construction.

Huge building losses take place in lighter construction because of high winds (tornadoes and especially typhoons and hurricanes). Great damage is done by water (i.e. rain) when the building envelope is breached by winds. This is often due to shortage of fasteners and emphasizes the need for a robust envelope.

Chapter

4

The Hazard Scenario

Circumstances which require robustness are those where critical physical conditions exceed the limits of resistance which the structure was equipped with, in terms of strength, deformability, durability, etc., or where the real resistance of the structure is lower than anticipated, or a combination of both.

The essence of a *hazard scenario* is that the resistance of the structure has been overcome, leaving it in an impaired, damaged or altered state. Examples for this are very diverse and include rupture of certain elements in a system, yielding, instability, displacement, reduction in cross section, etc. Large deformations often accompany the event, which must be considered when analysing the altered structure.

It is perhaps useful to classify the events in a very general sense since different *families* of scenarios require different approaches to robustness. We shall call one family the *interior flaws* or simply *flaws* where the origin of the event is located within the structural system. The second family will then be *exterior causes*. Forensic investigation of accidents often finds that a combination of causes relating to both classes is responsible for the mishap, i.e. a weakened structure was subjected to loads that exceeded the design loads. It may be that one of two or several events would not have been sufficient to cause distress and only their cumulation did.

4.1 Interior flaws and the like

Structural properties such as strength, stability, stiffness, durability, etc. are variable quantities. For the design and analysis of a structure, one substitutes hypothetical values for the real ones, which can be determined only when the structure exists (this may be difficult even then, involving complicated or indirect testing procedures). A variation will therefore exist between the hypothetical and real values. This variation has been the subject of much research and a large amount of probabilistic as well as forensic data exist comparing *specified* (or *expected*) values with tests or reality. These data were used for the establishment of safety factors (among other considerations) and a certain amount of variation is considered *normal, usual* or *legitimate*, i.e. not much can reasonably be done to reduce it. The variation follows more or less a Gaussian distribution for random data, with some special considerations for extreme deviations–which one tries to eliminate through testing, inspection or quality control procedures, recognizing that large variations are less *legitimate*, even though probabilistic theory permits them to exist.

Philosophically, it is easy to see why this is so, for three reasons:

Firstly, large variations of the structural resistance from *nominal* or *mean* values are more frequent in reality than the theoretical (Gaussian, etc.) distributions would permit. In other words, "they are less rare than they ought to be".

Secondly, variations in the properties of the product of a process guided or performed by humans are a consequence of the actions of the humans involved, including all kinds of errors which may infest those actions, due to lack of attention, communication, etc. This is borne out by the fact that where a large variation of properties is found, usually someone responsible for it can be identified, having committed a human error.

The third reason for considering large variations as illegitimate is the fact that the larger the variation, the more visible and obvious it is usually to observers who can apprehend it and initiate a correction process. Research has shown that gross (human) errors in the creation of (design) information are frighteningly frequent initially. Most errors will however be eliminated subsequently through a sort of filter process before they can take effect. This filter process takes place when the original data become translated into different forms (calculations into drawings, drawings into fabrication and placing processes, etc.). During this process, the information is seen and reviewed by a number of individuals who will catch oddities and irregularities in the pattern, or will apply targeted checking procedures. However, even after the filtering process, human errors remain the principal cause of large and fatal variations in the properties of structural systems. By and large, human errors are responsible for a real frequency of structural failures that exceeds the theoretical one (based on legitimate probabilistic variation) by several orders of magnitude. Human error is therefore very much an element of the flaw scenario.

What this means is that, due to the nature of gross human errors, which initially can take any form or magnitude, the filtering process is of great importance. It will eliminate most of the obvious and more massive ones, leaving some of lesser magnitude. No probability law has been found so far for gross human errors but, similar to the legitimate deviations, the smaller ones will be more numerous than the larger ones. Smaller ones may however combine with other events to cause critical scenarios. The consequences of this can only be compensated by robustness.

4.2 External causes

The list of external events that might affect structural systems to the point where *design resistance* is exceeded and robustness must be relied upon is long and of a great variety.

It includes such things as gas explosions or vehicle impact, the probability and intensity of which can be reasonably assessed based on past experience. It also includes the effect of extreme natural disasters, for which data again exists permitting predictions of probability vs. intensity–although manifest and dramatic surprises are still possible, as the tsunami of December 2004 has demonstrated. It includes the actions of terrorists, which are difficult to predict, as recent experience shows, too. It is difficult as well with anthropogenic conditions, as they will largely depend on future technology, economy and the interaction of nature with man's behaviour.

What is common to most of this is that the intensity of the event that must be expected can vary within a very wide range. Sometimes some parameters of a *maximum credible event* can be estimated based on known physical limits, such as for instance the maximum mass of a truck. Already the speed of the run-away truck is much more difficult to assess; in addition, the energy of the impact will be proportional to the square of that speed, making things even less certain. It is similar with the ground motion caused by earthquakes, the intensity of which follows a very steep curve where with increasing return period (rarity) the amount of energy released may vary through several orders of magnitude.

Most of the time, the effects of external events are expressed in terms of forces or deformations that the structure will experience. For the assessment of robustness it is often of great importance which of the two is predominantly imposed onto the structure:

- A run-away truck will impose a force, an impact and a quantity of energy onto the structure it is colliding with.
- A foundation movement (which might be considered an internal flaw if the structural system is to include the soil next to the foundation) will impose deformations.
- An earthquake will impose dynamic displacements, to be amplified or attenuated by the structure, accompanied by the generation of internal forces.

If information on such an external event is sufficient, it can be made an element of the ordinary design of the structure, to be analysed in the usual way, through mathematical or physical modelling. This will normally result in a structure that is strong enough to resist the effects of the event without losing its function.

In some cases this may turn out too costly and where safe upper limits cannot be established for the effects of the event, other ways must be found to establish structural survival. Present day design of earthquake-resistant structures is one example for this where an amount of damage in terms of permanent deformation, cracking, etc. is deemed acceptable and compatible with the function of certain structures.

For the assessment of external loads and events mathematical or physical/experimental modelling is used much in the same way as for the assessment of the structural properties. Logically therefore, the numbers used in the design and analysis are subject to the same influence of human errors as was discussed in the context of flaws, the only difference being that the other side of the safety expression is now affected.

$$\underset{\text{(prediction subject to error)}}{\text{Resistance}} > \underset{\text{(prediction subject to error)}}{\text{Effects of exposure}} \tag{4.1}$$

This introduces situations where the event escapes description at least in the quantitative sense. It brings us back, as in the case of the unknown flaw, to the domain of truly unpredictable events for which structural robustness of the structural system in a general sense must be sought, designing it with a topology, geometry, material properties, dimensions or auxiliary elements, etc., so that the structure will survive the loss or reduced resistance of any or all of its elements.

4.3 The consequence

The task must now be turned around to examine each and every element of the structure, its role in the structural system and what the description and the consequences of its failure would be, regardless of the (unknown) scenario, which might attend to this. It starts to resemble the situation of a living being that must be prepared to deal with any and every circumstance, in order to protect is own survival. This has been the *real life* situation of all structures from time immemorial, and a kind of evolution through *survival of the fittest* can be seen to have taken place over time, favouring good and robust kinds of construction, sending others to oblivion.

Chapter

5 Considerations on Hierarchies

5.1 The hierarchy of failure modes, targeting quality control

Most structures can fail in different ways with each mode of failure answering a description of the course of events, as well as of the consequences for the system; two common examples will illustrate this:

- The failure in compression of a column in a building is usually sudden, with little advance warning, and in most cases will be of catastrophic consequences, especially if it supports several floor plates.
- The failure of a beam in bending will often be paramount to yielding of steel only, which will cause permanent deformations, and redistribution of loads but not much else. Beams usually carry only one section of a floor so that the scope of a beam failure is limited to that particular section.

Robust structures are those that develop less catastrophic failure modes first which very often leads to a subsequent limitation of forces associated with the event (see Section 6.7: *Capacity Design and the Fuse Element*) or to remedial action. It is possible in design and analysis to rank the failure modes of a structure in terms of a hierarchy in such a way that the less harmful ones are generated at lower loading levels. This must be done without load and resistance factors, with realistic values representing the physical conditions and on a probabilistic basis. Usually one introduces sufficiently high ratios between the load intensities belonging to different failure modes; typically they will amount to factors of 1.3 or so, in order to ascertain that uncertainties of modelling or material properties (e.g. overstrength), etc. cannot upset the assumed hierarchy in the real structure.

In earthquake engineering this has for example led to the *weak beam/strong column* concept where one designs the columns to resist higher load intensities than the beams which can be designed for ductile behaviour much easier and more effectively.

A very important corollary of structural hierarchy concerns the quality control. Usually the means available for the detection of flaws and errors in terms of time, money and logistics are not sufficient to exercise exhaustive control and verification of each and every element of a structure being planned, fabricated and erected. The question then becomes: Where do we wish the limited means to come to bear, in order to catch as much as possible of the consequences of Murphy's Law[2] before they can take effect?

This is an optimization problem in the sense of maximizing the effect of one's investment. A number of aspects and ramifications need to be considered which will help to decide the strategy for quality control, some of which are not directly related to robustness such as timing, choice of personnel, methods of verification, testing and inspection, repetitiveness, etc. [10-13]. One of them, however, follows directly and trivially, and one might say, from considerations of structural hierarchy: It will be a good idea to concentrate attention and means on the elements of the primary structure, and in particular those which are prominent on the main load paths and/or sensitive for the quality and care of execution: Main connections, such as weldments, bolting assemblies, anchorages, as well as those elements which are in a position -topologically or through their properties- that they might trigger progressive failure.

5.2 A possible classification of methods

Although explicit risk analysis is not the method of the subsequent consideration and risks will not be quantified at all, respective thinking can help to classify possible methods of increasing the robustness of structures. A risk R can be expressed by the following term [6]:

$$R = \sum_{i=1}^{N_H} p(H_i) \sum_{j=1}^{N_D} \sum_{k=1}^{N_S} p(D_j \mid H_i) p(S_k \mid D_j) C(S_k) \tag{5.1}$$

with the denominations: N_H is the number of hazards H_i, N_D the number of direct (local) damages D_j, N_S the number of types of follow-up behaviour S_k, $p(H_i)$ the probability of occurrence of hazard H_i (first term), $p(D_j|H_i)$ the probability of the occurrence of direct damage D_j due to hazard H_i (second term), $p(S_k|D_j)$ the probability of the occurrence of structural behaviour S_k due to direct damage D_j (third term) and $C(S_k)$ the (monetarized) consequences of structural behaviour S_k (fourth term).

Methods that try to reduce the probability of occurrence of hazards, i.e. the first term of Equation (5.1), can be summarized as *event control* (EC).

To minimize the probability of a local damage due to a hazard, i.e. the second term of Equation (5.1) is known as the *specific load resistance method* (SLR). One possibility to minimize the probability of progressive collapse in the case of a local damage, i.e. the third term of Equation (5.1) is the *alternate load path method* (AP). These two methods are called *direct methods* or *direct design approaches*.

Indirect methods or *indirect design approaches* provide minimum levels of strength, continuity and ductility and reduce both the second and the third terms.

Finally, the *reduction of consequences* tries to minimize the fourth term by appropriate provisions.

Obviously the following Chapter 6 could follow the sequence of methods described above. The reasons why this scheme is not followed are:

[2]Edward Aloysius Murphy Jr. 1918–1990: *"Whatever can go wrong, will go wrong."*

- Some aspects raised in Chapter 6 cover more than one method.
- Other aspects are treated in a controversial manner, i.e. measures that exclude each other are discussed.
- Human errors are not reflected in this scheme although they are paramount reason to require robustness.
- Some aspects may be objectively or subjectively unknown at the stage where Equation (5.1) is applied. In other words: Both the various probabilities $p(\ldots)$ and the numbers N_H, N_D and N_S may be subject to major changes during the lifetime of a structure.

Prescriptive documents such as codes usually ignore that latter fact, because nobody can be forced to know what cannot be known at the time, where decisions have to be made; textbooks are freer in this respect. Nevertheless, in Section 6.17 the covered elements of robustness will be integrated in the scheme described above.

Chapter

6

Elements of Robustness

The menu includes a number of strategies and considerations, some of which may not be mutually exclusive but can or must be combined or belong to more than one course in the menu:

- Strength
- Structural integrity and solidarization
- Second line of defence
- Multiple load path or redundancy
- Ductility versus brittle failure
- Progressive failure versus the zipper stopper
- Capacity design and the fuse element
- Sacrificial and protective devices
- The knock-out scenario
- Stiffness considerations
- The benefits of strain hardening
- Post-buckling resistance
- Warning, active intervention and rescue
- Testing
- Monitoring, quality control, correction and prevention
- Mechanical devices

Each of them may be applicable in certain circumstances but not in others. In particular, the type of event and the type of structural response will decide about the strategy of choice:

- Is the event load controlled or deformation controlled?
- Is it repetitive?
- What are its physical limits in terms of forces, impact, energy, deformation, movement, duration, etc.?
- What will be the conditions following the event? Structural function, further events?

6.1 Strength

To provide strength beyond the theoretically needed minimum is often the only applicable as well as the most economic strategy to enhance robustness. Wherever the use of brittle materials or situations cannot be avoided, e.g. slender members in compression, those elements must be protected from overload, often through extra strength. Building codes are recognizing this by assigning lower material or member strength reduction factors to such elements for purposes of structural design, sometimes hidden in the alchemy of code writing. These rules are usually written with common types of structural systems in mind. In the case of large or novel structures, it is left to the designer to create a rationale for safety margins and for the members and failure modes.

To equip critical elements of a structural system with extra strength is a practice dating back to prehistoric times, in all cultures, with the corner stone symbolizing the quest for robustness and the great pillars of the historic cathedrals being just two of the expressions of this principle. As it appeared over time that even the mass of the main piers did not prevent worrisome signs of distress, one made them even more massive by adding buttresses, as in the case of the Hagia Sophia whose proportions then inspired the construction of the great mosques of Istanbul.

In modern structural design, the grading of relative strength and the provision of supplementary resistance to critical elements form the basis of capacity design and analysis.

6.2 Structural integrity and solidarization

The term *structural integrity* has been used in codes and literature rather freely, without much attention to what, in every context, was meant. There is, however, a family of circumstances where structural integrity is of an entirely physical significance.

Many structures, especially older works, are constructed simply by placing elements one on top of the other so that gravity loads can be transmitted through bearing contact. Horizontal forces are ignored or assumed to be sufficiently small not to overcome *stiction* (i.e. adherence through friction) at interfaces. This has turned out to be a major source of risk with many dramatic collapses being caused by elements leaving their supports due to seismic shaking, or long-term movements. It is paramount to a lack of structural integrity, and consequently of robustness in the literal sense since the structure, if not tied together, risks disintegrating into its parts which are allowed to move relative to one another.

Classic design and analysis assume that every structure, if not separated intentionally by expansion joints, will act as a unit, differences in displacement of the various parts made up exclusively by (small) deformations of the elements. Mainly in the context of seismic resistance, the validity of this hypothesis has come into question, and the analysis of overall behaviour, floor plates and roof structures acting as diaphragms is recognizing that problems may exist here, after all, and need to be looked after in the rule book. The (especially dynamic) analysis of systems with soft diaphragm elements turns out to be very difficult and results are questionable; an incentive exists therefore to make everything as rigidly coherent as feasible, in the model as well as in reality. This in turn makes it necessary to also provide the strength throughout the system to resist the forces that will act in consequence, especially at the connections in an assembly of prefabricated elements.

The aspect of structural integrity and coherence can also, in a different context, be interpreted as *solidarization* where one ascertains that one individual member cannot be destroyed all by itself but a number of elements are acting together to provide sufficient resistance. An example for this is the collision strut that has been used in construction over railroad tracks: Sufficient resistance is provided by the compound resistance of a series of columns.

6.3 Second line of defence

Some structures have different ways in which they can resist loading. The situations which come up most often in this context are those illustrated earlier with the guardrail along roadways: If the limit of resistance of a member in bending is reached due to loads acting perpendicular to its axis, it will deform through a kinematic mechanism involving a certain number of plastic hinges. These, in the case of ductile elements, permit large deformations through yielding of the material, which changes the geometry of the member. It can now act in a hammock-style fashion in tension; it has found a second load path, which offers much more resistance in terms of keeping the colliding vehicle on the road. Of course, it does not provide a very pleasing sight after the accident and the administration of public works will have it replaced in time. However, it has given a robust performance by fulfilling its main function.

Similarly, building frames can be designed in such a way that the loss of one column does not lead to widespread collapse, with the floor plates having deflected and acting in a hammock-like way. This is not done very often, as it implies substantial additional cost among other reasons. Usually, it is more economical to provide extra strength to the columns where they are exposed to a perceived hazard such as vehicle collision etc.

If the *second line of defence* strategy is used, the analysis of the structure in its deformed/damaged condition must be carried out with care, representing as closely as possible the anticipated scenario:

If a beam or plate must convert to a hammock-like load path, its continuity and strength through all connections and elements on the load path must be ascertained accordingly. Numerous cases are known where a structure would presumably have resisted collapse, if it had not been for the connections that failed prematurely, or because some other elements became overloaded in the process, such as for example the remaining columns in the frame.

6.4 Multiple load path or redundancy

This proposition resembles the *second line of defence*, the only difference being that several different load paths are *engaged* or *mobilized* initially, with forces being channelled through all of them. If one or more of them fail, the rest may be able to continue resisting the loads, but a number of conditions must be met for this to work:

- The remaining structural elements must be strong enough, collectively, to resist the loads corresponding to the situation after the event, which was associated with the failure.
- Members becoming overloaded in the course of the event must be able to deform without losing their resistance, so that other load paths (members) catch up with sharing the load. If they cannot do so, a zipper-style failure may ensue! (See following discussion).

- The remaining load paths must include the auxiliary structural elements transferring the load from the failed elements to the alternate load path.
- The remainder of the system must substitute all structural functions of the failed elements, e.g. conserving overall stability.

A prominent example for a redundant structure is the wooden joist floor where most of the time, all conditions listed above are satisfied:

- Wood joists are currently designed for effective safety factors of 2 to 3 or so, depending on the exact definition. Two joists located either side of a broken one can therefore pick up its share of the loading. Matters will get more serious where more than one adjacent joists have lost their resistance (wood is essentially a brittle material, its resistance disappears irreversibly when overloaded). However, before this happens, joists undergo large, non-linear deformations.
- The floorboards that span perpendicular to the joists will normally perform the transfer of load from the broken joist to the adjacent ones, with a typical spacing from 300 to 600 mm.
- To prevent lateral buckling of the joists, they are normally interconnected by short pieces of boards or *St. Andrew's crosses* in every space between. These will still be functional following rupture of a joist.

All conditions for robustness through redundancy are therefore satisfied in this case and wooden joist floors are indeed a very robust type of structure as witnessed by the very large quantity in existence today (in North America alone, of the order of 10^{10} m^2 of wooden joist floors are found).

An extremely important caveat must always be kept in mind when redundancy is to be used to create robustness:

Multiple load paths involving brittle situations or limited deformability may lead to premature, zipper-style failures. The importance of this point deserves some illustration:

Examples for the zipper style progressive failure are found in a number of catastrophic accidents involving heavy suspended ceilings. One of them became notorious as it involved corrosion of stainless steel rods (Uster Switzerland [16]). A heavy cement-based ceiling was suspended over a swimming pool, with a large number of so-called (as we know now) stainless steel rods. These rods had become weakened and embrittled over time, due to corrosion, which had eaten into the section of the rods, promoted by the chlorine and humidity of the air of the pool, which circulated in the space above the ceiling. Not only was the corrosion nearly invisible (the hangers had been inspected shortly before the accident), but also it was much localized so that the weakened steel could no longer yield to produce any appreciable extension – a case of brittle elements. The accident had all the characteristics of a progressive failure involving a large number of parallel elements (i.e. load paths). It started with one of the rods rupturing which created a new loading situation where another rod became overloaded and ruptured and another and another. . . The collapse happened so quickly that a number of swimmers did not make their escape and were crushed under the weight of the fallen debris. A highly redundant structure had failed progressively, due to brittleness of its members. Similar accidents keep being reported.

Had the rods been ductile, the rupture of one of them would very likely not have caused a real accident since redistribution of loads and safety margins would have been sufficient to stop the collapse.

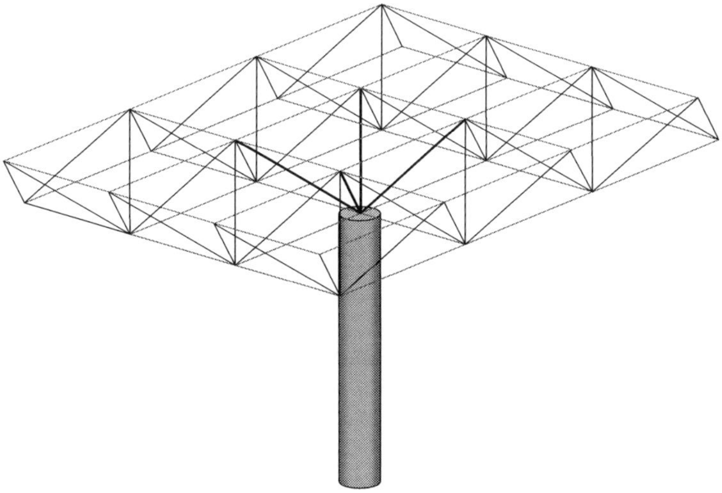

Fig. 6.1: Typical configuration of a space truss (the principal diagonal members are shown with fat lines)

Consider also the popular *space frame* or better *space truss*. It is usually supported on a series of points with a number of diagonals transferring the loads directly to the supporting columns (*Figure 6.1*), therefore a situation of redundancy exists – the highly hyperstatic structure providing as many different load paths as there are diagonals connected to the support.

Space trusses are usually composed of slender members (pipes, rods) with bolted or welded connections. Compression members will be of a brittle-type behaviour (loss of resistance after overloading, through buckling), whereas tension members are themselves not brittle and would be able to plastify, if it were not for the connections. Those are usually less resistant than the member itself, due to weakening from bolt holes, welds, threads cut into the section or because the connection piece is less resistant than the members attached to it. In all these cases, the connections will rupture before the yield strength of the members is reached. This amounts to a brittle behaviour once more and the redundancy provided by the several load paths is of a negative value (see also Section 9.10: *Space trusses* and Section 10.6: *Power lines*).

6.5 Ductility versus brittle failure

As we have seen, ductility is essential to the beneficial effect of redundancy, second line of defence and similar approaches to robustness. Its definition has now been agreed upon fairly precisely for practical considerations, although in some cases, rather arbitrary assumptions must be introduced. By and large, ductility is the ratio

$$\text{Ductility} = \frac{\text{Maximum deformation}}{\text{Maximum elastic deformation}} \tag{6.1}$$

where the term *maximum elastic* may sometimes be a matter of debate, especially for materials or situations where there is a gradual decay of stiffness, stiffness being defined as

$$\text{Stiffness} = \frac{\text{Load increment}}{\text{Deformation increment}} \tag{6.2}$$

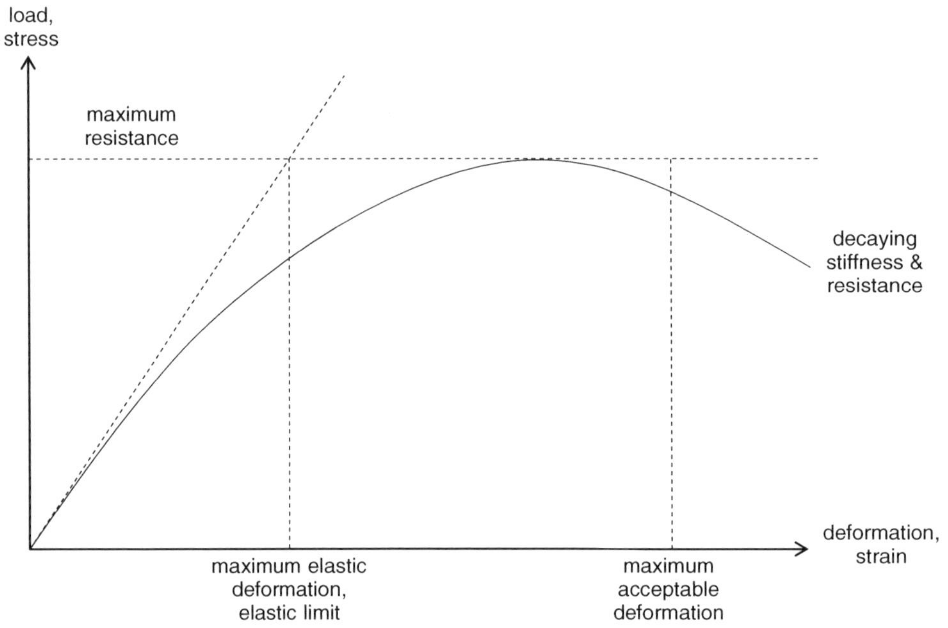

Fig. 6.2: Typical behaviour of ductile elements

Examples for this abound and one usually determines a virtual point of the elastic limit by intersecting two straight lines in the diagram describing material properties, member or system behaviour, one of them representing behaviour at low amplitudes of deformation, the other the horizon of maximum resistance, at constant loading (see *Figure 6.2*).

The *maximum acceptable deformation* is equally variable and its definition/description will follow the circumstances of each particular scenario. Perhaps the largest common denominator, encompassing all or almost all cases, can be given by the following verbal expression: Maximum deformation which the system can tolerate maintaining a useful resistance; where *useful* would be determined from the general definition of robustness in every case. This logically leads to an implicit relationship, quantitatively in the design for robustness, and therefore to an iterative approach, to be resolved through trial-and-error procedures, illustrated in diagrammatic form in *Figure 6.3*.

Most structural systems are of the character suggested by the curve in the above diagrams where resistance, after its maximum value has been reached, will gradually reduce, due to second-order effects, cumulative damage, reduction of the resisting critical cross section, etc.

Exceptions are certain metals or alloys such as *mild steel* which exhibit *strain hardening* beyond the yield stress (which in this case is taken as *maximum resistance,* see *Figure 6.4*) which can be exploited for robustness (see Section 6.11: *The Benefits of Strain Hardening*).

Similar exceptions concern steel-reinforced elastomeric bearings, which are behaving in a strain-hardening style until bond fails at the interfaces elastomer/reinforcing or the metal-reinforcing ruptures. After this has occurred, the bearing no longer functions in the originally intended way but will permit the portion of the structure above to settle onto its support which

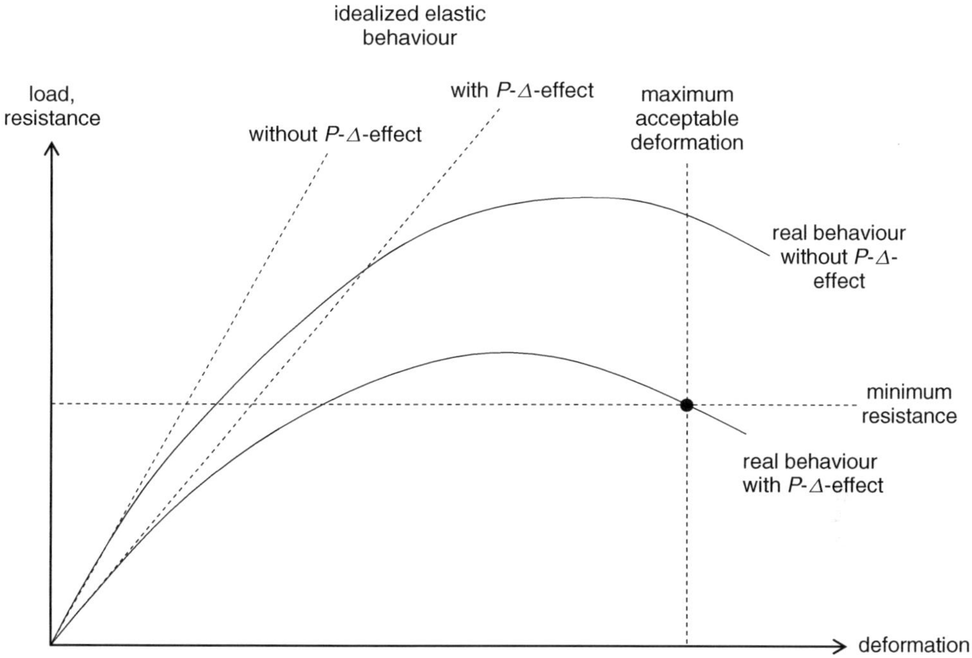

Fig. 6.3: Design for sufficient ductility

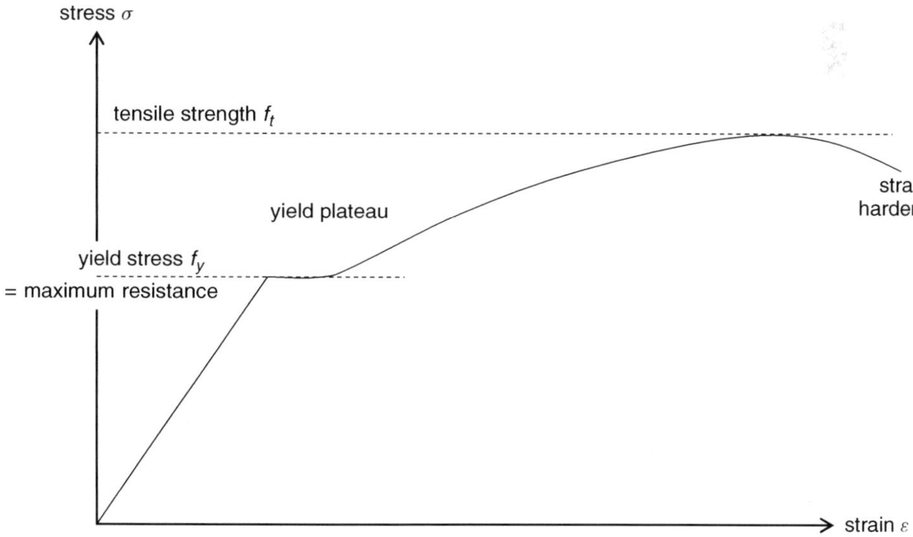

Fig. 6.4: Strain-hardening behaviour of metals

may or may not constitute its loss or failure. Failures of elastomeric bearings are known to have occurred without causing any further distress, meaning that the robustness of the overall system was not affected.

Brittle (i.e. non-ductile) behaviour can be the result of a number of causes that are listed subsequently.

6.5.1 Material properties

Different materials have different degrees of brittleness (or inversely, of ductility), roughly in the order indicated in *Table 6.1*.

Building material	**Ductility**
Mild steel, low alloy aluminium	More than 10
High-strength steel, metals	5–20
Reinforced concrete	1–10
3 Wood	1–3
Stone, brick, terracotta	1–2
Unreinforced, low-strength concrete	1–2
Unreinforced high-strength concrete	1–1.5
Glass, ceramic, stone, most fabrics	About 1

Table 6.1: Range of ductility values for common building materials

6.5.2 Local weakening (holes, notches, damage corrosion, etc.)

In bolted or riveted steel assemblies one often finds that, although all the materials are of a ductile nature, the system will fail in a sudden fashion because of the weakening at the bolt holes. It has been found that this effect can sometimes be compensated by the strain-hardening effect of good-quality (mild) steels, but higher-strength steels do not usually have properties permitting this. The basic consideration (for an element in direct uniaxial tension) is

$$\frac{\text{Section at bolt holes}}{\text{Total section (no holes)}} > \frac{\text{Yield strength}}{\text{Tensile strength}} \tag{6.3}$$

Notches or local corrosion may have similar effects with yielding even of a ductile material limited to small region, preventing the deformations necessary for redistribution of forces to occur prior to fracture.

6.5.3 Weak connections (weaker than the adjacent elements)

Nailed assemblies of wood or similar materials behave in a ductile fashion although wood is by itself a relatively brittle material. In this case, it is the nailed connections, which provide the deformability/ductility.

6.5.4 Fatigue

Fatigue under repeated loading cycles involving tensile stress will reduce ductility of even the best materials, as well as their strength. This phenomenon is well known and must be accounted for in all situations involving numerous loading cycles ($>10^3$ or so). It is usually not of importance in cases of one time peak load (collision impact, explosion, earthquake, etc.) unless the loading history includes stress variations producing cumulative fatigue damage.

6.5.5 Instability due to slenderness

Slender compression members fail generally in a brittle fashion, depending on the rate of slenderness. Exceptions of this rule can be found where buckling is of a local nature only, so that the portion of the element which does not take part in the buckling still possesses sufficient resistance for robustness (see Section 6.12: *Post-Buckling Resistance*).

6.6 Progressive failure versus the zipper stopper

A considerable variety of structural systems belong to a class, the characteristics of which make them prone to a particular type of progressive failure which has been nicknamed *zipper effect* or *domino effect*. Some of the most spectacular structural failures belong to this category, as for example the famous *liberty ships* of the Second World War, or more recently, the collapse of entire power lines up to 100-km long in the region of Montreal, Canada, where hundreds of lattice towers carrying the main (750 kV) supply for a large part of the land, fell under the weight of ice which formed on the wires, in domino-like fashion (see Section 10.6: *Power lines*). Large rips in fabric roofs belong to the same class, most fabrics used for roofs being of a brittle type (glass, aramid fibres) as for example in the roof of the Montreal Olympic Stadium which failed twice, once due to wind and once due to snow accumulation and ponding (see Section 9.13: *Fabric structures*). In many cases the use of materials or elements with brittle response characteristics is the only choice as in the two latter examples mentioned above. Hydroelectric towers in inaccessible places must be light and easily assembled as well as tall and are therefore composed of slender members with little residual post-buckling resistance. Fabrics for tensile roofs where the fabric acts as primary or secondary structural elements must be strong as well as stiff and resistant to creep deformation. No fabric has been invented to date that has all these properties to a sufficient degree as well as being sufficiently ductile and cheap.

The answer to the problem of proneness to *zipper type* progressive failure is found in the basic goal of design for robustness: To limit the extent of the failure: Its nickname is the *zipper stopper* where in a group or sequence of brittle elements, periodically *strong points* are created which will stop the progression of the failure. In the case of long welds unravelling, cross welds, stiffeners, gaps or bolted transitions may be arranged. For the power line, every fifth or tenth tower or so may be of a stronger or more ductile type, with possibly a certain trade-off between strength and ductility, depending on the precise characteristics of the scenario being considered.

For the fabric roof, a two-level structural hierarchy may be used with a mesh of crossing *strong bands* or strengthening cables, possibly with anisotropic properties, and a lighter, isotropic in-fill of the mesh, the assembly being sewn/glued/welded together in place.

6.7 Capacity design and the fuse element

A *fuse element* in the structural application differs from the well-known electrical fuse which interrupts the transmission of current completely when overloaded. A *structural fuse* is an element which maintains the transfer of load at a certain level (usually through ductility) thereby limiting the forces which are being transmitted through the load path of which the fuse is an element. This is very useful in all deformation-controlled situations such as imposed displacements (e.g. through foundation movements or the effects of strong earthquakes).

Of course, the deformability of every fuse element is limited and robustness requires that this limit is not exceeded. In most circumstances the level of resistance and stiffness is not maintained at precisely the same level throughout the entire range of deformations of interest but varies to some degree, either upwards as in the context of strain hardening of steel, or downwards, as for instance exhibited by reinforced concrete elements, in particular when exposed to cyclic load reversal. In modern texts one talks about ductility being maintained when resistance is not degraded more than 20% or so.

Fuse protection also requires that the entire system still maintains its resistance with the fuse element having been deformed – in some cases, structural systems may fail through excessive deformation (e.g. second-order effects affecting stability) before the limit of deformability of the fuse elements is reached.

Usually, deformations beyond the limit of elasticity are permanent, at least partially. In the case of severe earthquakes, deformations are of a random nature, as a consequence of the dynamic response, i.e. their magnitude can be determined only probabilistically. It may therefore be necessary to purposely limit these deformations so that the structure will not become a loss because it cannot be redressed at reasonable expense (see Section 6.10: *Stiffness Considerations* for discussion of this aspect).

Non-linear deformations (i.e. those deformations exceeding the elastic limit) are usually associated with substantial *dissipation of energy*, which escapes in the form of heat. Energy dissipation can be represented graphically through hysteresis diagrams, or its rate is quantified as damping. Where deformations are reversible or cyclic, as in seismic situations, one must reckon with some alteration in the material properties in the parts being deformed non-linearly which often amounts to a degradation of resistance and stiffness compared with the initial (first cycle) maximum. This may not be of serious concern in some situations such as rare and severe earthquakes (which will take the fuse through very few non-linear cycles) but becomes increasingly critical with increasing numbers of load reversals or cycles leading to brittle behaviour (see Section 6.5.4, discussion of *fatigue*). The fuse strategy must therefore be limited to events which produce few excursions into the non-linear range.

In a fuse-controlled system, forces will be positively limited but deformations will not. Analysis of such behaviour must therefore include the rest of the structural system when it is being deformed beyond the elastic limit (or maximum resistance) of the fuse element. Some other elements may be deformed beyond their maximum resistance in turn and must be designed for sufficient deformability. In addition, progressive deformation in the sense of a cumulative "ratchet" effect may be the consequence of cyclic loading beyond the elastic limit.

6.8 Sacrificial and protective devices

In some cases, it may be uneconomical or impossible to provide sufficient robustness in a structural element, or its description and behaviour may not lend itself to such a strategy. It is then useful to consider the introduction of devices that will intercept the impact or forces constituting the danger scenario such as in the case of building columns exposed to potential vehicle impact. Rather than making the structural column strong enough to sustain resistance to the impact of a runaway truck or a derailed railroad engine, it may be cheaper and more effective to build a sufficiently massive separate structure between the source of impact and the structure to be protected. Damage or destruction of the protective device may be acceptable in extreme and commensurately rare situations, but the locomotive cannot be allowed to roll over it. For the design of such devices, one may have recourse to either the law of conservation of energy or of momentum, depending on the type of information at hand.

This strategy is particularly suited to situations where the loading scenario cannot be determined with any useful certainty. This is the case for structures erected in water, deep enough to permit large ships or icebergs to travel. The colossal mass of these objects makes it difficult and uneconomical to build structures, the main purpose of which is something else (e.g. as oil platforms, bridges), to resist impact. Protective structures such as artificial islands may provide an answer. They will either resist or deflect the impacting vessel – or iceberg – away from the vulnerable structure itself, which can now be designed for wind, waves and seismic forces only [1].

For the vehicle impact scenario, one can often make reasonable hypotheses for some of the parameters (mass, speed, length of crushing zone in the vehicle, etc.). The design of the structure is usually based on the acting forces, which can be obtained in very simple fashion, as an average for the duration of the impact, because it can be quantified either by the momentum or by the kinetic energy (Equations (6.4) and (6.5)).

$$\text{Force} = \frac{\text{Momentum}}{\text{Duration of impact}} = \frac{\text{Kinetic energy}}{\text{Length of crushing zone}} \tag{6.4}$$

or

$$F = \frac{mv}{t} = \frac{mv^2}{2l} \tag{6.5}$$

with m and v denoting the mass and the speed of the vehicle, respectively, l the length of the crushing zone of either the vehicle or the sacrificial structure alone, or both combined and as a result the duration t of the impact is found.

$$t = 2\,l/v \tag{6.6}$$

This assumes a constant rate of deceleration. Tests have shown that this is not strictly the case but that for sufficiently deformable structures, the resistance (reaction) always corresponds to the force acting, and the use of average values leads to reasonably safe designs.

6.9 The knock-out scenario

Cases are found where it would be impractical or impossible to provide sufficient strength or ductility, and the possibility for a structural element to be accidentally removed must be considered. To provide robustness, a second line of defence type of strategy may be chosen, or a particular element may be designed to be knocked out, without causing serious consequences.

Examples for this family of scenarios are mostly related to impact through collisions or explosions: Building columns within reach of vehicle impact or elements enclosing storage of potentially explosive substances[3], where in some cases, knock-out panels in walls or roofs are intentionally provided, much in the sense of a fuse to protect the rest of the structural system by letting the pressure escape.

Some considerations should be remembered in these contexts, which, although they may appear trivial, are nevertheless vital for the establishment of robustness:

- The knock-out element must really be *knocked out*, i.e. it must not *hang on* to the rest of the structure, dragging it along when it leaves its position. This can be achieved for example by providing, intentionally, a brittle element on the load path which will break with certainty at a force level which all other elements can safely resist.
- The knock-out element must not become a projectile, endangering other nearby systems. This can be achieved for example by providing an articulation (real or plastic), which allows the knock-out element to open without flying away.
- The remainder of the structure must remain in a place and in a condition that is compatible with its function/survival/robustness.
- In the case of collision, it is easy to see that an isolated column is more vulnerable than a wall of some length. For an explosion, the opposite is true – knock-out panels for explosions should therefore present a sufficient surface to the explosive pressure to provide relief and cede quickly enough and at a pressure level sufficiently low for the rest of the envelope (roof, floor, walls) to resist.

6.10 Stiffness considerations

In many cases, deformations must be limited as they themselves may jeopardize robustness, being the cause for loss of the structure because of collapse through instability (e.g. second-order effects like the $P - \Delta$-effect) or because the deformed structure can no longer serve its purpose and must be taken down, rehabilitation being too costly.

Classically, deformations are inversely proportional to stiffness. Exceptions are deformation-controlled situations such as the case of seismic effects since the dynamic response of the structure is mainly related to the amplitude of the ground motion. Still, a minimum stiffness must be provided, be it only to produce acceptable performance under wind loads (i.e. residual loads in the sense of the principal expression of robustness, see Equation (3.1)).

Stiffness, at force levels relevant to robustness criteria, is a variable quantity, which in general degenerates with increasing loads and loading cycles. This is of particular interest in cases where several parallel load paths exist in a system, with different stiffness properties.

A well-known example for this are the mixed shear wall/frame structures intended to resist seismic effects which were quite popular in the recent past, so much so that building codes devoted special chapters to their design.

[3]This case is treated in much detail in Ref. [15].

The idea is to create two lines of defence, a first one in the form of stiff (especially in terms of shear) shear walls of limited ductility, the second one a moment-resisting frame with enhanced ductility but lesser overall shear stiffness. Using elastic analysis one finds an interesting interaction of the two elements, with a shear reversal in the upper part of the structure which can even be demonstrated algebraically.

The problem with these systems is the large difference in overall shear stiffness:

$$\text{Shear stiffness of shear wall} \gg \text{Shear stiffness of frame} \tag{6.7}$$

Looking at practical cases – many high-rise building designs followed this concept in the 1960–1980s – it is found that often an unfavourable course of events will develop at high loading, with the shear wall resisting most of the loads until its capacity is reached and its resistance lost quickly. This leaves the frame on its own. Although it must usually be designed to provide the strength for 25% of the calculated base shear it will often possess a stiffness that is much smaller in proportion and be too flexible to effectively resist this loading, and the structure will deform excessively and risk to fail because of the $P - \Delta$-effects (*Figure 6.5*).

This aspect was left out in many designs, as the building codes did not address it. The concept has now been eliminated from some of them.

One can of course design frames which offer sufficient stiffness (comparable with the stiffness of a wall), but one usually finds that the dimensions needed for the beams and columns to do this, are too massive to be compatible with architectural and economic considerations.

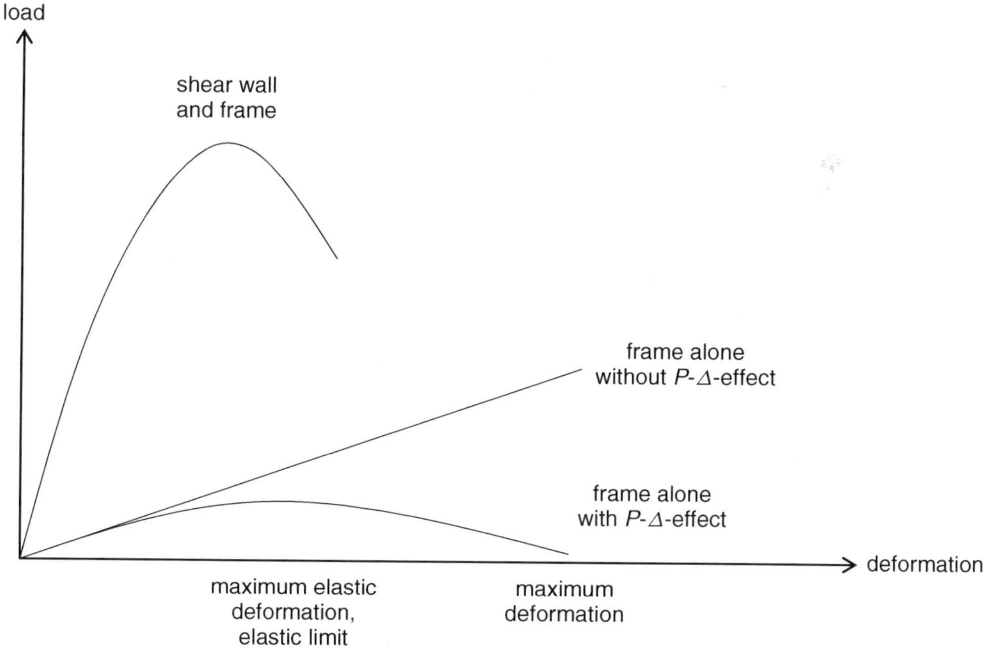

Fig. 6.5: Behaviour of a combined system consisting of shear walls and moment-resisting frames

In many recent earthquakes, it was found that buildings with variable stiffness along their height have fared badly. This is because in a displacement-controlled situation, most deformations that must occur to accommodate the imposed displacements will concentrate in the most flexible portions of the structure. They are likely to cause local failures in these locations as can be seen in numerous photographs of buildings which have lost their base storey because it was a *soft storey* which could not accommodate the large deformations imposed onto it.

The fact that imposed deformations accumulate in the zones of minimum stiffness can be observed in a wide variety of situations. For example:

Continuous concrete floor plates must accommodate shrinkage effects (the invention of concrete without shrinkage is still outstanding despite all claims to the contrary). They are likely to shrink, if restrained on two or more points, at the location where their stiffness is reduced, e.g. by an opening, or by a crack which may have formed for another reason. The tensile stiffness across the crack is reduced to that of the reinforcing steel alone which will now be stretched by the tension resulting from restrained shrinkage; this will cause the crack to open wider and the floor plate must deflect in order to re-establish contact in the compression zone, to re-establish equilibrium for positive bending moments (see *Figure 6.6*).

One can interpret this as a lack of robustness as this type of deflection may be quite substantial, putting the function of the floor in question.

A similar case is found to exist in bolted connections of steel members. Bolts require holes in the pieces to be connected which constitute a local weakening of these elements, in terms of stiffness as well as strength. If exposed to tensile strains beyond the yield strength, the critical (weakened) section plastic (i.e. yielding) deformations will occur exclusively at these locations, with a very limited length of material participating, all the other portions of the structure being still in their elastic state, contributing little to the overall deformation. The deformability (plastic range) of the zone at the bolt holes is therefore quite small and if substantial deformations are imposed (i.e. demand) on the system, it will reach its limit strain quickly and rupture in tension.

In order to eliminate the problem, one can of course reinforce the zone of the bolt holes by thickening the plate, but this is rather onerous and may introduce other problems

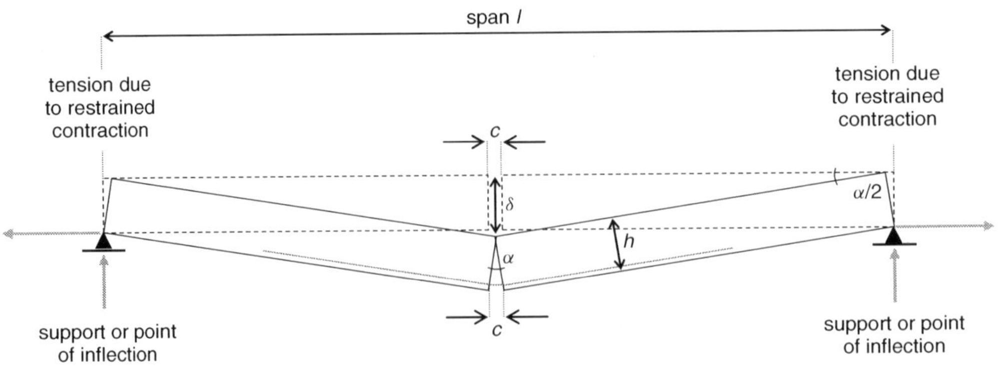

Fig. 6.6: Deflection due to restrained contraction

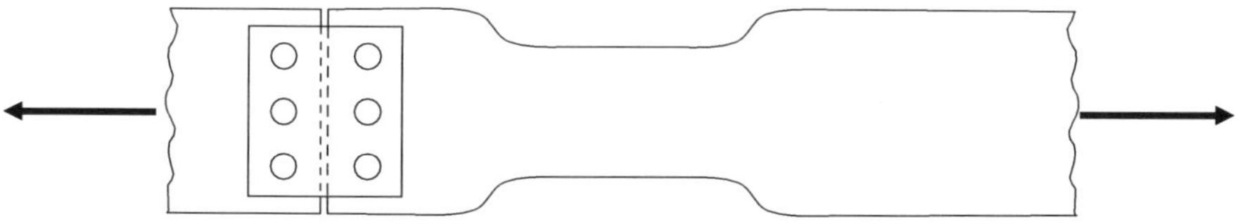

Fig. 6.7: Transfer of plastification away from weakened section

(e.g. associated with the welding of reinforcement plates). Another possibility is the artificial weakening (narrowing) of another zone on the same load path nearby but away from the bolt holes (see *Figure 6.7*) displacing the zone of plastification to a location where a sufficient length of material can be mobilized to participate in the yielding (see also Sections 6.5.2, 6.5.3 and 6.7).

For metals with a sufficient strain hardening, it is found that the problem can be solved without such measures (see Section 6.11: *The Benefits of Strain Hardening*).

Another way to deal with the problem merits to be mentioned: It is to remove the critical (less ductile) zone from the place of maximum stress. Often this is not difficult and quite obvious, as for instance in the case of beams exposed to variable shear and bending where the locations where plastic hinges may form can be anticipated so that bolted connections can be placed away from these.

6.11 The benefits of strain hardening

Materials and elements exist that exhibit an increase of stiffness with increasing deformation (strain). In the case of steel, this is called *strain hardening*. Traditionally, it is most pronounced in the so-called *mild steels*, i.e. steels with low or modest yield strength (see *Figure 6.4*).

Modern steels of higher strength are being offered today with similar properties at substantial premiums. However, most of higher (yield) strength steels being sold derive their yield strength from mechanical/thermal treatment whereby the material was already taken into the range of strain hardening and has therefore a distinctly different behaviour (see *Figure 6.8*).

If deformability is critical, this type of steel should be used with caution. Consider the case of the bolt holes: For a steel with pronounced strain hardening and a tensile strength exceeding the yield strength by a sufficient margin, it has been found that the weakening at the bolt holes may become less (or not) critical since:

$$\begin{pmatrix}\text{Tensile strength}\\ \text{at bolt hole section}\end{pmatrix} > \begin{pmatrix}\text{Yield strength}\\ \text{elsewhere}\end{pmatrix} \tag{6.8}$$

In other words: Yielding will take place elsewhere in the structure before rupture at the weakened zone. For pieces in direct (and constant) tension along the load path, the following condition must be fulfilled if rupture at low deformation rates is to be avoided. Equation (6.8) can be rewritten as:

$$\begin{pmatrix}\text{Yield}\\ \text{strength}\end{pmatrix} \times \begin{pmatrix}\text{Full}\\ \text{section}\end{pmatrix} < \begin{pmatrix}\text{Tensile}\\ \text{strength}\end{pmatrix} \times \begin{pmatrix}\text{Reduced}\\ \text{section}\end{pmatrix} \tag{6.9}$$

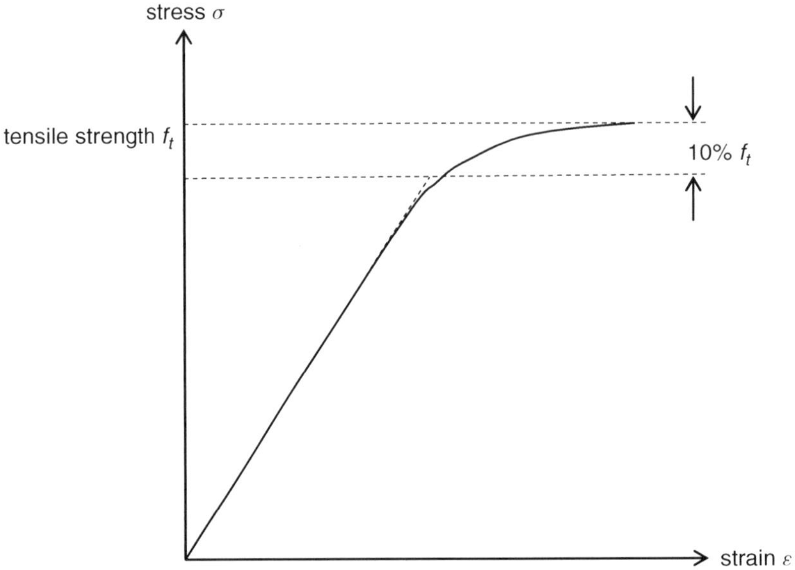

Fig. 6.8: Typical characteristics of higher-strength steels

Usually bolt holes are reducing a section by 20% or so. Consequently, if this is the case, the tensile strength of the material must exceed its yield strength by the same amount.

$$\frac{\text{Tensile strength}}{\text{Yield strength}} > 1.2 \tag{6.10}$$

For many modern high-strength steels, this cannot be guaranteed where tensile strength exceeds yield strength by only 10 to 15% or even less.

Another case where strain-hardening behaviour comes in handy is elastomeric bearings mentioned earlier. Their resistance (stiffness) increases with loading, the limit being effectively related to the tensile strength of the steel plates.

More generally, the benefits of strain hardening are quite obvious: Reserve strength exists which can be mobilized before deformations become excessive or rupture occurs. This is particularly interesting in the case of chain type (i.e. isostatic) systems or part systems: Before rupture is imminent, substantial deformations will take place, providing visible evidence that a problem of overloading exists, in other words: a warning.

6.12 Post-buckling resistance

In some instances, it is advantageous to provide relatively thin-walled elements in places where they will be exposed to compressive forces causing local or partial instability at loading levels below the limit state[4] (cylindrical structures, masts, airplane bodies, etc.). This was recognized

[4] The definition of limit state may vary from case to case and from country to country. It is used here in a purely qualitative sense, designating loading values which are used in strength design, together with limit state (reduced from real or specified) values for resistance.

some time ago and considerable research has shown that partial buckling may be permissible provided that the residual (post-buckling) strength is sufficient to resist the (limit state) loads acting. In a simplified approach to proportioning of thin-walled metal members, one removes the portion of the section that is permitted to buckle (i.e. evade the stress) for the calculation of structural resistance (see *Figure 6.9*).

A note of caution merits to be remembered here: Buckling is associated with important local deformations, which may cause rupture in cases of repetitive loading, through fatigue or cyclic yielding.

Formally, the exploitation of post-buckling reserve can be seen as a case of second line of defence which, however, does not seem to be of any practical significance, at least for the simple example mentioned above.

Another situation may be mentioned which is strictly speaking a case of post-buckling resistance. It is the well-known tension bracing where slender members are arranged in such a way that at least one of them works in tension while the remainder may buckle in compression when the loading exceeds their resistance. Due to the great slenderness which is typical of these braces – usually flat plates, small angles, rods or even cables are used – the buckling load is quite low and in the calculations compression resistance is normally not accounted for.

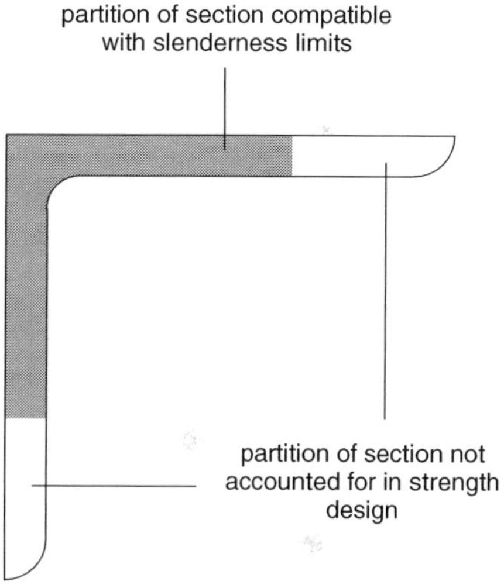

Fig. 6.9: Post-buckling resistance.

The robustness of these systems is questionable in certain cases, in spite of the – usually – ductile nature of the tension brace. The reason for this is the dramatic decay of stiffness of the structural system, which occurs as the structure has been deformed beyond the elastic limit in either direction. There will now be a range of displacements where no bracing is acting because the buckling resistance is negligibly small due to slenderness, and the tensile resistance will set in only after the previously produced plastic deformation is equalled again. This state of affairs can be described in a hysteresis graph shown in *Figure 6.10*.

It is quite unacceptable as it permits the structure to reel uncontrollably, between two limits where the structure will receive a shock every time they are reached. It is not compatible with robustness in most cases and tension braces will have to be designed not to exceed the elastic limit.

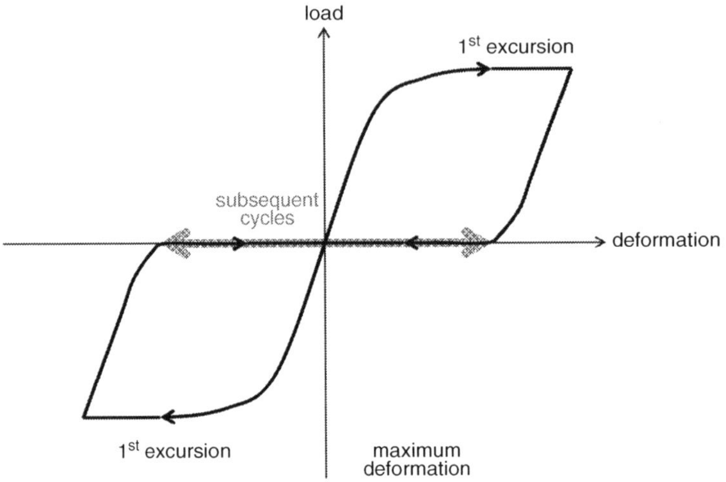

Fig. 6.10: Behaviour of tension braces

6.13 Warning, active intervention and rescue

Evacuation and emergency measures such as auxiliary supports, reduction of loading, etc. have long been used as means to evade danger and risk. All of these take their time to become effective which is the time between the danger manifesting itself through perceptible signs until evacuation is complete, or until the auxiliary system has become operational.

This consideration is not limited to dangers and risks for life and limb but applies with the same logic to other cases of distress which may be related to the loss of a structure when it is found beyond repair, which state of affairs might have been prevented had the relief been introduced in time. In these situations, survival of the structure with its function intact (which is generally the essence of robustness) is a question of timely and effective human intervention and robustness becomes a question of it being enacted.

The enactment of emergency measures in turn is dependent on the awareness that an imminent danger exists, as well as on the time and means available to put them in place.

In many circumstances, this is found to be the case to a sufficient measure, or through design it can be brought about. Where it is not, a serious problem with robustness exists! Warning signs and the behaviour of the structural system up to loss of function or rupture, as well as the correct interpretation of these are elements of robustness here, and it is worthwhile to reflect on ways to make them work, and to identify cases where and why they might not.

Warning signs are physical phenomena, which are paramount to, precede or accompany the beginning of distress of the structure. Most of them will be of a visual nature, sometimes not perceptible to the naked eye but only through instrumentation. Some may be perceptible to anybody, others only to the specialist. Some may be hidden from view, or may manifest themselves through *proxy effects* such as jamming doors, etc. An enumeration of phenomena that can be indicators of a structural problem cannot be exhaustive for all situations, so the following checklist will only cover the symptoms more frequently seen:

- cracks in the structure itself, or in architectural finishes, separation of contiguous parts, splitting, tearing
- deformations or displacements, out-of-alignment, out-of-plumbness, buckling, bulging of walls
- spalling, scaling
- corrosion, rot, chemical degradation, mouldy smell
- jamming, absence of movement where it was part of the design
- excessive vibration, change in frequency of sound
- noises
- looseness of fasteners, missing fasteners
- leakage, water infiltration, discoloration
- fretting, loss of tension, kinking of wire strands or cables
- abrasion, loss of section, etc.

Each situation will have its typical indicators, which are related to its particular character: Materials, organization of structural system, exposure, environment, history, attention to quality and maintenance. Some of them may be harmless although they look dramatic to the lay observer. Others may look innocent but represent signs of imminent distress.

If human intervention is made an element of systemic robustness, it must be adapted to each situation. In some cases, a structure may show signs the interpretation of which is uncertain, even to the specialist. The usual answer to the question: "What to do?" is to install a protocol of monitoring in order to observe progression or its absence, of the signs of possible distress. One would like to avoid situations where "the cure turns out to be worse than the illness", because intervention was premature or misdirected. In other cases, immediate action must be undertaken, as for example in the case of a structure showing signs of serious degradation, with a winter with probable heavy snowfall imminent. In this particular case, two options may be possible, each with his drawbacks: Reinforcing or shoring of the structure, which may be costly and obstructive, or snow removal – where the mobilization of workers may be difficult and uncertain.

In any case, wherever human intervention is made an element of robustness, possible human shortcomings must be accounted for: Inattention, lack of communication, absence – it is difficult to organize a 24-h watch – incompetence, malfunction of equipment and instrumentation etc.

These are serious considerations especially in situations where time is short so that only quick and well-directed action can keep things from getting out of hand.

For certain typical scenarios of emergency or distress, human intervention is institutionally organized as in the case of firefighting, power failure and hurricanes, the reason being that these situations are recurring with sufficient frequency and with a typical description to warrant substantial investment into the organization of protective or corrective action on a stand-by basis. Failures of structural systems – fortunately, one might say – do not often belong to this class and because of their great variety, they do not lend themselves to permanent installation of emergency response organisms. It is therefore difficult to conceive of adequate intervention to deal with short-term emergency situations of structural distress or imminent failure since the time is often too short, and other strategies likely will prove to be more adequate. In situations where the problem takes longer to develop (weeks, months, years), this is different of course, when there is sufficient time to perceive, observe, confirm, interpret and prepare adequate measures to deal with the situation while it develops. The description of the scenario must therefore include the element of time.

Human intervention can also be directed towards the exposure of the structural system to loads or other environmental effects. An example for this would be snow removal or active snow melting, or strictly speaking, the installation of auxiliary protective devices (see Section 6.8). Other features belonging to this category are indications of load limits, through signs or obstructions limiting the size or weight of vehicles accessing; their effectiveness must always be evaluated with a view on the future reality; e.g. the sign limiting floor loading must not be painted over, removed or ignored.

Other examples include shoring or otherwise stabilizing structures in distress, e.g. following a destructive earthquake.

6.14 Testing

Numerous situations exist where the representation of a structure through mathematical or physical models is uncertain and one would like to have confirmation of structural behaviour beyond calculations using such models.

It can be stated that the more extreme or innovative the situation of a structure becomes, the greater will be the uncertainty of results from analysis, i.e. the greater must be the difference to be expected and accounted for, between reality and its desk top representation. Traditionally, physical testing of representative samples or of the structure itself has been called upon in such cases to help bridge the gap.

There is a basic problem with this: In order to learn something useful about the structural behaviour in an extreme state, the scenario would have to be simulated in the test, i.e. in essence a destructive test. This, for industrial mass and serial production can be and is being done routinely with prototypes and randomly selected samples. For many structural systems a prototype for testing purposes is out of the question, every structure being essentially one of a kind and representing a single investment of such magnitude that its destruction, partial or entire, cannot be afforded.

For these practical reasons, structural testing for robustness is therefore limited to samples of repetitive elements rather than entire systems, or to non-destructive methods, both of which are not directly telling us what we wish to know.

As a typical case for this, laboratory representations of single elements out of large structural systems are tested in isolation and subjected to rates of deformation which they would never see in reality since those deformations would long before have led to the collapse of the structure due to second-order effects and such.

Typically then one finds that ductility and conservation of stiffness are less than perfect and do not correspond to the classic bilinear graph one has learned to retain. For reinforced concrete, in particular, it is virtually impossible to achieve perfect ductility, and a more relaxed definition of this property must be adopted in order to avoid excessively complicated reinforcing which is difficult to place and therefore difficult to control.

It is never a good idea to rely on complex construction methods in an environment of cheap materials and high competition which most construction is.

6.15 Monitoring, quality control, correction and prevention

In Section 6.13, responsive human intervention was discussed. Human intervention can also be preventive and by the same logic, be interpreted as an element of robustness.

In the form of inspection, verification, quality control or assurance, non-destructive or proxy testing, it will be directed mostly towards the elimination of defects, deficiencies, inadequacies, all of which are often the result of shortcomings of the people involved in the creation of a structural system. Discovery of an anomaly, of something which is amiss, wrong or suspicious, is at the start of the process of correction which takes place mostly before the structural system is turned over to use and intended function. It must be followed by interpretation, confirmation and corrective action. As a rule, the later this takes place in the production process; the greater will be the cost of correction. This might suggest that to be cost effective, such efforts should be concentrated at the front end of the construction process, i.e. in the design stage where it is cheap to correct shortcomings. This will not help of course against errors and defects which are introduced later in the process, and it is easy to see that the optimization of the effort towards the elimination of errors becomes quite involved [10-13]. Discussion of these questions in the past has produced some insight of qualitative nature, and has triggered a great worldwide effort under the heading of quality assurance (ISO 9000 Code series), which is still in the process of being enacted. Unfortunately, it is being misdirected towards institutionalized paperwork and the production of vast amounts of documentation, which turns out to be quite ineffective with respect to the quality of construction, i.e. the elimination of defects and shortcomings.

Likewise, monitoring through periodic inspection does not provide automatic assurance: The symptoms of real distress such as telltale cracks[5] may be evident only to personnel with sufficient specialized knowledge or experience whose alertness has not become dulled over time. Recent events and developments have shown that especially large organizations such as railroad systems or public highway departments have fallen victim to management imposed budget constraints, cutting back on frequency and thoroughness of inspection, employing less qualified personnel and delaying or neglecting follow-up.

Automatic monitoring is sometimes considered attractive to control risks one suspects to be inherent in a structural system. Experience with this sort of approach has been mixed to say the least. Typically, periodic readings are produced, involving advanced electronics for sensors, data processing, display and the triggering of alarms. Reliability of highly technical systems in the real world remains an elusive goal, however. This has a number of causes:

- Sophisticated equipment needs careful and frequent maintenance itself.

---

[5]Concrete structures very often show a variety of cracks most of which do not directly affect structural performance. It may therefore be difficult to identify the ones which are of real concern.

- Electronics are being developed and tested in clean and controlled environments; when exposed to the reality of construction or the environment, they often break down or malfunction.
- Electronic monitoring devices work with low voltage and currents. Nearby power lines, transformers, etc. often produce secondary currents which are more powerful and may cause short-circuits, burnouts and other types of malfunction in the sensitive monitoring apparatus. Lightning may do the same.
- Monitoring devices are often installed at construction time – one wants to know about the behaviour of the structure as early as possible. Handling of heavy mass materials, power tools and construction machinery by uninformed workers creates an environment that is very unfavourable to the functioning and survival of sensitive equipment.
- Data retrieval and digestion require informed and educated personnel, which may be available at the start but will be replaced in time, the specialized know-how getting progressively lost.
- The expenditures related to monitoring are not usually producing any revenue and will therefore, modern accounting being what it is, be unpopular with management and therefore on a low and likely decreasing budget.

All of this must be kept in mind if one wishes to delegate the maintenance and assurance of robustness to the function of automatic apparatus, or to human intervention and informed watchfulness. Again: A structure in distress will not provide a friendly environment for complex and sophisticated activities, or: When the chips are down, the reliability of such systems may fall victim to the same events the structure does – a very real positive correlation exists for this.

6.16 Mechanical devices

For the control of movement or forces or to dissipate energy, mechanical devices have become increasingly popular in certain situations. They can be grouped in two classes:

- Passive devices that do not require energy input.
- Active devices the function of which depends on a – most of the time electrical – power source.

Considerations of reliability almost invariably lead to conclusions favouring passive installations for reasons already discussed in the context of human intervention: Will the device on which survival of the structural system depends, work when matters get serious as for instance during an extreme wind storm or in a severe earthquake when power grids are likely to fail and humans will rather be looking after their own safety than the functioning of the back-up power source? In general, therefore, one prefers to entrust structural safety and survival to mechanisms such as tuned mass dampers, friction or sloshing dampers, which will absorb energy, converting it in heat when and where the environment feeds it into the structure.

Even passive devices need to be looked after to keep them functional as corrosion or other types of chemical alteration may take their toll, jamming parts that are meant to be moving when the time comes. And again: The more complex and sophisticated the apparatus, the more likely it is to become dysfunctional, usually with a definitely positive correlation to extreme events, i.e. if matters get serious, the likelihood of a failure will increase.

6.17 Summary

The different elements of robustness can be tabulated according to their function described in Section 5.2 (*Table 6.2*).

Elements of robustness	Event control	Direct approaches		Indirect approaches	Reduction of consequences
		Specific load resistance method	Alternate path method		
6.1 Strength		x		x	
6.2 Structural integrity and solidarization			x	x	
6.3 Second line of defence			x	x	x
6.4 Multiple load path or redundancy			x		
6.5 Ductility versus brittle failure				x	
6.6 Progressive failure versus the zipper stopper					x
6.7 Capacity design and the fuse element					x
6.8 Sacrificial and protective devices					x
6.9 The knock-out scenario					x
6.10 Stiffness considerations			x	x	
6.11 The benefits of strain hardening		x	x		
6.12 Post-buckling resistance			x	x	
6.13 Warning, active intervention and rescue	x				x
6.14 Testing		x	x		
6.15 Monitoring, quality control, correction and prevention	x				x
6.16 Mechanical devices	x				

Table 6.2: Covered elements of robustness ranged according to methods based on risk analysis.

Chapter

7 Maintaining Robustness

Older structures have usually seen a history of alteration, modification, remodelling, redecoration, etc., which leaves them in a state that may be quite different from the original and may no longer conform to the hypotheses, assumptions, specifications and plans which formed the basis of the design and construction. Often, these alterations (holes, notches, cuts, transfers, etc.) are of a minor nature when seen individually and have been introduced at different times, for variable reasons and by different agents. Their sum, however, may be quite significant, and usually of a negative nature, i.e. reducing such desirable structural properties as strength, resistance, stiffness and robustness. Finishes, hiding the messy character as well as the weakening represented from view, usually cover much of the modification. It is a good idea to consider this *fact of life* and if one wishes to depend on certain structural elements and their resistance, to verify if they can really be counted on, or if the safety margins their original design included, has been eaten away even where no signs of distress are detectable.

The popular argument that a structure that has stood the test of time, having survived for a century, will do so for another 100 years is entirely fallacious unless supported by sufficient evidence that the original state of the system is still intact. There is no reason to assume that a building will respond differently to use and abuse than a motorcar that, after it has travelled a certain distance, needs to be repaired, nursed, rebuilt or replaced. There is no reason to believe either that our ancestors built more solidly, with more attention to quality and robustness than we do today. The *good old times* never were, and the plague of giving all work to the low bidder has been with us since the beginning of time, as witnessed for instance by the famous letter Vauban[6] wrote to his superior Louvois[7] on July 17, 1685, concerning work done by the low bidder and by contractors forced to provide rebates (*Figure 7.1*).

Since the document shown in *Figure 7.1* is written in French, a free translation by the first author follows:

"A number of works are waiting which are not completed and never will be. All this is due to the confusion which is caused by the frequent rebates which are made on our works, for it is certain that all these broken contracts and promises, re-adjudications are only attracting

[6]Sébastien le Prestre de Vauban (1633–1707) was General Inspector of fortifications and Field Marshal to King Louis XIV of France in numerous wars. He is also the author of scientific work on taxes and statistics.

[7]François Michel Le Tellier, Marquis de Louvois (1641–1691). War Minister to King Louis XIV of France.

the scum among the contractors, the misfits and the ignoramuses whereas the ones who are serious and capable of conducting business are running away.

Moreover, I am saying that this is delaying the works as well as raising their cost and spoiling the execution, for these rebates and successful negotiations are imaginary; as a man who is about to drown hangs on to anything he can grab, the contractor will not pay his suppliers and workers, he will steal from whomever he can and employ only the worst because they are cheaper. He will also use only the worst quality materials, try to fight about everything and scream at everyone for pity.

Enough of this My Lord, to make you understand the shortcomings of this conduct. For God's sake re-establish good faith, pay what the works are costing and do not refuse to give honest reward to a contractor who has fulfilled his duties. This will always be the best deal you can find."

The conclusion from this is clearly that robustness cannot be assumed to persist on its own, but must be reassessed and re-established if one wishes to count on it.

Cette lettre, écrite par Vauban
le 17 juillet 1685 à Louvois,
conserve, deux siècles et demi plus tard,
toute son actualité.
Le prix de vente au plus fort rabais
est la plus mauvaise de toutes les solutions !!

Il y a quelques queues d'ouvrages des années dernières
qui ne sont point finies et qui ne finiront point, et tout cela, Monseigneur, par
la confusion que causent les fréquents rabais qui se font dans vos
ouvrages, car il est certain que toutes ces ruptures de marchés, manque-
ments de paroles et renouvellements d'adjudications ne servent à
vous attirer comme entrepreneurs tous les misérables qui ne savent où
donner de la tête, les fripons et les ignorants, et à faire fuir tous
ceux qui ont de quoi et qui sont capables de conduire une entreprise.
Je dis de plus qu'elles retardent et renchérissent considérablement
les ouvrages, qui n'en sont que plus mauvais car ces rabais et bons
marchés tant recherchés sont imaginaires, d'autant qu'il est d'un
entrepreneur qui perd comme d'un homme qui se noie, qui se prend
à tout ce qu'il peut, or, se prendre à tout ce qu'on peut en matière
d'entrepreneur, c'est ne pas payer les marchands chez qui il prend
les matériaux, mal payer les ouvriers qu'il emploie, friponner tous ceux
qu'il peut, n'avoir que les plus mauvais parce qu'ils se donnent à meilleur
marché que les autres, n'employer que les plus méchants matériaux,
chicaner sur toutes choses et toujours crier miséricorde contre
celui-ci et celui-là.

« En voilà assez, Monseigneur pour vous faire voir l'imperfec-
tion de cette conduite, quittez-la donc et au nom de Dieu rétablissez
la bonne foi : donner le prix des ouvrages et ne refusez pas un honnête
salaire à un entrepreneur qui s'acquittera de son devoir, ce sera
toujours le meilleur marché que vous puissiez trouver. »

CHANTREAU ET CIE, *IMPRIMEURS A NANTES*

Fig. 7.1: Vauban's letter to Louvois (1685)

Chapter

8 Conclusion

The toolbox for dealing with the problem of robustness is endowed with numerous strategies, which in many circumstances leaves several options open to choice. This of course constitutes a classical optimization problem: Where do I put my money and effort so it produces a maximum effect. Because quite often, numerical assessment of these effects, in terms of probabilistic or deterministic quantification, is out of reach because of the high rates of uncertainty inherent in the scenarios affecting robustness, and evaluation will remain qualitative and subjective. This may not be such a bad thing as it permits to bring fuzzy-type information and subjective experience to bear, which does not depend on scientific "rules of the game" and availability of significant data, to be considered valid.

Chapter

9

General Applications

9.1 Punching failure of flat plates (strength, ductility, second line of defence)

This is a type of failure which, in spite of being well known, even notorious, still occurs with frightening frequency. A number of reasons can be cited for this to happen, not least the mere volume and complexity of current building codes[8] which have become an important source of errors themselves as design engineers are forced to delegate their task to the *black box* of the computer.

The mechanism of punching shear failure is well known and does not need to be discussed here. One feature however is important to note in the context of design for robustness: The real shear resistance of a concrete slab around a column head is a function of the tensile strength of "real life" (not laboratory prepared) concrete, among other effects such as cracking, eccentricities, etc. It is therefore highly variable. This fact puts the problem into the class of situations that are highly infested with uncertainty, this time affecting the resistance side of the expression for structural safety.

Three different approaches have been proposed and used to provide sufficient robustness in this case, and certain building codes are addressing some or all of them:

- Increase slab thickness and/or provide stronger concrete over the columns.
- Provide shear reinforcing in the critical zone.
- Provide bottom reinforcing through the column.

In this discussion, it is exemplarily easy and clean-cut to assign each approach to a certain *strategy for robustness*, with the different connotations and consequences of each one.

The first method is obviously to provide sufficient *strength*. Building codes provide criteria for it, which persist being the subject of never-ending discussions in the code committees about precisely how thick is thick enough. In some instances, regulations have been made stricter than in the past which raises the notorious problem of the existing construction, which was designed to less restrictive requirements that are presently considered insufficient. Does this construction now need to be condemned or rehabilitated?

---

[8]The International Building Code, published in 2003, has XII + 660 pages and weighs 1.7 kg. For the design of a concrete slab one also needs to study the ACI Code with 369 p., weighing 1.0 kg.

To provide a thicker slab also increases the cost of building on several counts and is in contradiction with architectural criteria, *mushroom slabs, drop panels* or *column capitals* being undesirable where the slab itself forms the ceiling.

The criteria for sufficient slab thickness are often quite complicated. They are relating to a very brittle mechanism of failure, which is also influenced by local forces and stresses due to eccentricity and bending. Some building codes are attempting to address these effects, resulting in rather complicated procedures.

The second method amounts to a change of the critical mechanism *from a brittle to a ductile mode of failure*. Various details have been proposed to provide shear reinforcing, some of them proprietary. The practicality of shear reinforcing, especially in thinner slabs, may become questionable, requiring watch maker precision to become effective, or making the exercise illusory where that cannot be consistently achieved.

The third approach is a good example for a *second line of defence*, which will become effective when a shear failure develops. The bottom reinforcing, which is anchored in the column, or passing through it, will act in a hammock-like fashion, keeping the slab from falling down (onto the slab below, triggering the collapse of that one and the ones below it in a progressive collapse as has occurred in some terrifying instances).

This cannot be achieved with the top reinforcing that will tend to peel off, unless it consists of bent bars, a style of reinforcing which was used in the past, but was all but abandoned because it required meticulous attention to the complicated placing sequence.

The bottom ("integrity") reinforcing is made mandatory in some building codes and has been shown to be effective in numerous tests. Considering the persistent notoriety of punching shear failures, it seems to be a good idea to invest in robustness in this way, in spite of the additional cost.

As a fourth approach one might mention structural steel inserts (shear heads), which act in a similar way as the shear reinforcing shifting the critical zone away from the column where shear stresses are sufficiently low for the unreinforced concrete section to resist. This method is favoured mostly in Europe as it allows a flat underside to relatively thin slabs, without the encumbrance of drop panels or capitals. The cost of shear heads is quite substantial, however, and fireproofing must usually be provided.

The choice of strategy obviously depends on the circumstances, i.e. the requirements or incentives of each particular case, which may relate to economics, architecture, tradition or preference of the respective designer. The important thing is that the punching failure mechanism is properly considered in the structural design. Recent failures provide the proof that this is not done consistently, out of ignorance, neglect of some other type of human error.

9.2 Unbonded prestress

Prestressing, or most often, post-tensioning with tendons embedded in grease and encased by ducts made from sheet metal or polymers has been a popular method of equipping concrete members, especially slabs in buildings, to resist bending moments. The theoretical basis for the design of such elements is well known, and the advantages of omitting the onerous and relatively messy operation of injecting cement grout into the ducts are easy to see.

Unbonded post-tensioning is still rigorously promoted; experience in practice though has been mixed to say the least, and in some cases, outright disastrous. One of the authors of this text remembers vividly a 33-storey office building which could only just be saved from having to be condemned and torn down because relaxed strands had been found in the slabs. Painstaking forensic detective work and extensive testing revealed that the relaxed strands were confined to one specific part of the structure along with a reason being identified why this was so (in the construction records it was found that the particular slab had been exposed to exterior conditions for a longer period of time than planned, due to an interruption of construction, having to do with financing problems).

The litany of potential and real problems related to unbonded post-tensioning is rather dramatic. Taken together, it is paramount to an absence of robustness of such structures, with respect to a number of common and trivial scenarios.

- If an unbonded tendon is weakened anywhere along its length, the entire unit has lost its function. This may happen through heating by a small fire, mishandling during construction, local corrosion, thoughtless drilling or cutting (electricians and plumbers have diamond drills which will cut through anything. . .), loss of bond at the anchorage, etc.
- Most vulnerable are the anchorage zones where water often finds a way to enter the duct, promoting corrosion – highly stressed steel is prone to stress-corrosion. These zones are particularly vulnerable to exterior conditions which, for high-rise buildings where the method is often used, may vary from warm humid summer to frigid winter conditions with all what this implies in terms of the presence and penetration of water. Until the structure is enclosed in the building envelope, it usually remains exposed to the weather for weeks or months.
- Failure of structures with unbonded post-tensioning can be sudden, without warning. In order to detect faulty (relaxed) tendons, the concrete structure and the duct must be opened to bare the tendon.
- Changes in loading patterns will cause the tendons to slide which may lead to corrosion promoted by friction.
- Modifications of structures with unbonded post-tensioning are very difficult if not impossible. The same holds true for strengthening when part of the post-tensioning is found to have been lost, or higher loads must be accommodated.
- The filling of the space between the strands and the duct is impossible to control. Empty spaces are prone to fill with water over time from condensation, migration of humidity in the concrete, water left over from construction time. Because of the difference in density with respect to the grease, the water will accumulate in the depressions of the ducts (positive moment region), displacing the grease.
- During construction, the ducts may get punctured, especially during the placing of the concrete. This will permit water entry directly, as well as air circulation, leading to corrosion.
- The grease does not provide any chemical or electrical protection (which cement grout is doing).

All of this put together makes for a scary catalogue of potential causes of failure, persuasive to some, but not to others, especially to younger engineers who have little experience with reality but faith in theory. It is a situation where Murphy's Law can easily take effect, and its consequences have been found very serious indeed on certain occasions.

Slabs with unbonded tendons are meant for ideal conditions as represented in textbooks or promotional material, where successes are exclusively represented. If unbonded post-tensioning must be used, all of the above-mentioned must be addressed and satisfactorily resolved.

9.3 High-rise buildings, high strength concrete, a difficult case

Intensive research, driven by the economics of urban real estate, has created the methods and procedures needed to produce concretes with compressive strengths which were not even dreamed of a few decades ago. Most of this new material is used for high-rise building columns and sometimes walls where economic pressure is very intense to minimize the footprint of vertical elements, in order to maximize rentable area. Other applications also exist but they are quantitatively less important.

High-strength concrete is a considerably less forgiving material than traditional concrete. It is relatively more brittle, i.e. more prone to sudden, almost explosive failure (did you ever watch a specimen being tested for compressive strength of, for example, 80 or 100 MPa concrete?).

Considerations of robustness must then account for this fact, in the context of relevant scenarios:

- Seismic events.
- Sabotage.
- Human error, e.g. in the form of a bad batch of concrete which went unnoticed, or was discovered too late, or in the form of inappropriate modelling of the structure for the computer analysis which served as a basis for dimensioning.

Among the possible strategies to respond to this design challenge, i.e. to endow the structural system with sufficient robustness in spite of the unfavourable properties of the high-strength concrete, three approaches shall be briefly reviewed:

- to provide supplementary strength
- to reduce effective brittleness, i.e. to provide some ductility
- to provide a second line of defence.

9.3.1 The strength approach

Modern building code committees have been forced to consider the advent of high-strength concrete as a commonly used material, and they have, if hesitantly or reluctantly, done so by adjusting the design rules, mostly opting for leaning towards the safe side, i.e. being conservative.

In other words, safety margins were increased, in fact punishing the use of high-strength concrete, which amounts to an increase of the strength requirements, relative to traditional criteria for normal-strength concrete. The extra strength thus required is introduced in various more or less transparent ways into the algebraic expressions representing design criteria and – it must be said at this time – without much supporting evidence from research and practice – whether it is appropriate quantitatively what is being stipulated. Robustness is difficult to quantify, and even more so if it must be introduced into general design rules applying to a great variety of cases/scenarios. At least, code writers have recognized and acknowledged that a problem with robustness exists in the context of high-strength concrete, and have addressed it within the limitations of current knowledge.

Quantitatively, even the relatively more conservative stipulations of building codes with regard to high-strength concrete (from 10 to 20% with respect to traditional concrete) seem to be of little significance in the face of the increased brittleness (reduced ductility and deformability) of this material, and it is therefore worthwhile to look at other ways to bring about robustness.

9.3.2 Improved ductility

Economic incentives being very strong, high-strength concrete is also being used in regions of high seismic activity where the emphasis on material properties is concentrated mainly on deformability, the determination of realistic design forces being difficult and uncertain. Deformability of the concrete is a key ingredient in the design of ductile structures even though one tries to accommodate deformations in places where plastification of steel can be mobilized. Recent research has been intense on this subject, and the result is, rather persuasively, that the deformability of compressed high-strength concrete elements resides mainly in the amount and detailing of transverse reinforcement, which the codes now specify to be provided in amounts and degrees of complexity not seen before. The real cost of doing this is quite substantial in terms of money and time (= money again) and may offset the benefit of high-strength concrete use.

Where complex and cumbersome procedures are advocated in construction, the problem of reliability becomes more critical: Have the difficult reinforcing patterns been followed precisely in 100% of the locations, or have some of the more difficult-to-place reinforcing bars been left out in some instances, due to time pressure and less than effective supervision?

This would leave the structural system with *weak/soft spots*, which nature will readily find, according to fracture mechanics, as opportunities to concentrate deformations.

9.3.3 Second line of defence

In order to provide a second line of defence to compensate for the possible loss of a high-rise building column is a task for which theoretical solutions can obviously be found, in the form of hammock-style reinforcement of the floors, made from continuous elements of structural, reinforcing or post-tensioning steel, and accompanied by columns designed to pick-up load from a failed neighbour.

Practically, this may turn out to be less obvious, in terms of cost, as well as effectiveness (see Section 6.3: *Second Line of Defence*): If a floor plate or beam has to bridge over two spans, all elements affected must be equipped to resist the new forces: Columns will receive more load, slabs near column heads will be exposed to increased punching shear, horizontal elements will eventually be exposed to continuous direct tension, etc. In other words, redundancy that is equivalent to a second line of defence must be validated by demonstrating sufficient deformability of all elements. This may be difficult to achieve, and in the case of an exterior or corner column, it may be altogether impossible.

It may be concluded then that in the case of high-rise building columns, similar to numerous scenarios involving isostatic assemblies, no one-size-fits-all solution exists and the merits of each particular set of circumstances must be carefully assessed for the optimization of robustness.

9.4 The problem of the corner column (hierarchy of strength)

Classically, the corner stone was considered a very important and critical element of a building. Up into the twentieth century, it was put in its place by prime ministers, high priests or company presidents, for good reasons: Symbolically, one may say that its failure will affect two sides of the structure, from bottom to top, which event would, given the classic masonry construction, entail the collapse of at least a great portion of the building.

More modern types of construction have replaced masonry bearing walls and seemingly, this would change the situation through the use of more forgiving materials and different – improved – organizations of the structural concept. Surprisingly though, cases exist where similar considerations are the order of the day as they were for the corner stone, if not more so. One of these cases concerns the type of structure that relies on multiple frames near the skin of the building for the resistance to vertical and horizontal loads.

Consider a high-rise building with a square footprint 30 m on the side, and with 29 storeys. Each face of the structure is framed with ten columns with equal spacing, and spandrel beams of equal size at every floor plate (see *Figure 9.1(a)*).

Elastic analysis will show that vertical load is carried uniformly by all columns, except for the corner column which receives only half as much, or nearly so. Beams and columns will be engaged in the resistance to lateral loads, and the overall overturning moment will produce axial loads in the columns, according to their position (*Figure 9.1(b)*).

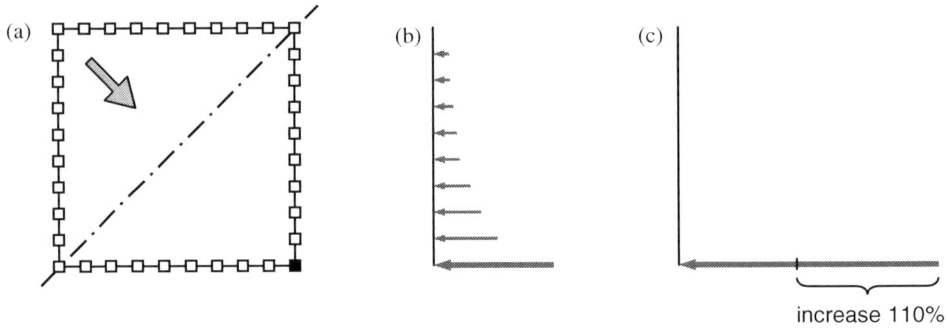

Fig. 9.1: Forces in columns of a high-rise building; (a) key plan. Forces in columns along a façade, (b) elastic force distribution, (c) plastic force distribution with strong column/weak beam design

Simplicity and architectural considerations will likely lead to a design where all beams and columns are of the same size and detailing, including the reinforcing steel in the case of a concrete structure, again with the exception of the corner column which will be designed to a different set of loading conditions, following the results of the elastic analysis.

So far so good, one might say, as the elastic design approach will also result in a *strong column/weak beam* structure as required/recommended for seismic design.

If one follows through and studies the mechanism which will develop at extreme loads, one finds a rather worrisome feature: When all or most of the weak beams are yielding, i.e. from plastic hinges at both ends, the sharing of the axial loads from the overall overturning moment changes dramatically: Where in the elastic analysis, most of the columns were participating, all of the overturning moment now ends up as axial load in the corner column. This effect will be especially dramatic when the lateral load acts in a diagonal direction (*Figure 9.1(c)*). It means in effect that the corner column is penalized by the inelastic behaviour of the structure, relatively to the other columns in the face of the building.

To make matters worse, several other features are adding to the predicament of the corner column:

- It will, if designed to the minimum requirements, as per building codes, be exposed to a lower stress due to permanent (gravity) loads and a load transfer due to creep will occur in time, shifting load to the corner column from its neighbours.
- A similar effect must be expected from the foundation where local soil pressure from permanent loads is likely to be lower under the corner column, providing the equivalent of a *hard spot*.
- The corner column has less reserve strength. If elastic analysis is used to dimension the corner column, the axial load from overturning effects will be more important with respect to gravity loads. The safety margins, which the building code stipulates via load and/or material factors, will be used up quicker than for the other columns when lateral load effects become extreme.
- Design for ductility is more difficult for the corner column than for the other columns as it is restrained less effectively against rotation and spalling, by the spandrel beams and floor structure.

None of this is considered in building code requirements, with the likely result that buildings of this type will fail progressively, starting with the loss of the corner column due to overloading in compression.

To alleviate the predicament of the corner column, two basic approaches are found on the menu:

- The flexural strength of one or several spans of the spandrel beam can be varied so that the last span provides a lesser resistance to bending due to shear from horizontal loads, thus reducing the potential overloading of the corner column.
- The corner column can be provided with extra strength over and above code requirements.

The first approach may lead to difficulties, where the letter of building code requirements based on elastic analysis must be satisfied, since the *strength hierarchy* among the spandrel beams and the columns will not be the same following the code rules as the one desirable for robustness. The second (strength) approach is more straightforward and simple as it will consist only in the *overdesign* of the corner column relative to minimum requirements of the building code.

9.5 Thermal deformations, compatibility and tolerance

All materials respond to temperature variation with deformations, which, if restrained, cause stresses which in turn may lead to damage and loss of strength and stability. Building elements are often found in situations where thermal deformations are not readily accommodated, meaning that partial or completely rigid restraints prevent expansion and contraction. Exposure to ambient conditions implies temperature variations due to contact with the ambient air, sun radiation, rain and snow, wind and accidental effects such as fires. Most of the climatic effects are cyclic with durations from seconds to hours, days or seasonal variations. This chapter will deal with building elements made from brittle materials that are restrained globally, i.e. at the extremities of the element, or locally, i.e. continuously.

Often, the origin of the restraint that causes thermal stresses is found within the element itself as for example in a wall, which experiences non-linear temperature gradients through its thickness, or in the case of a closed cylinder with a temperature gradient between outside and inside.

Exposure to the ambient conditions will cause the exposed materials to respond, starting at the surface, with the temperature variation penetrating into the interior from the exposed surface more or less quickly, depending on duration of the exposure, thermal mass and conductivity of the materials. This is not a stationary process and may imply some unexpected and seemingly counterintuitive symptoms, which in turn may lead to degradation and damage. Robustness in this context means that the structural system is tolerant with respect to thermal effects imposed by the environment, i.e. does not suffer undue or progressive damage, even over time. Temperature effects – expansion or contraction – are often superimposed on other, similar deformations due to shrinkage or creep, enhancing or counteracting each other: Concrete placed in the warm season will crack more than when it is placed in the late fall, since thermal deformation compounds with the shrinkage in the first case, and counteracts it in the second.

Most difficulties though arise from temperature differences or gradients among contiguous parts of buildings, which are continuously tied together, preventing differential movements. This may concern a masonry veneer, which is in contact – directly or indirectly – with the structure of the building. With its exterior surface the veneer will experience and follow the fluctuations of the ambient climate, while the structure remains at the constant interior temperature of the building being $20^\circ \pm 1^\circ$ to 2° C. If expansion and contraction joints are not provided in both directions, the veneer that is usually the lesser partner in terms of stiffness and strength may crack or buckle. This can readily be observed for example on numerous older and new masonry façades, where a vertical crack has formed near each corner: The veneer has formed its own joint of relief because the builders did not provide it. Often, these cracks are found to grow progressively over time in a ratchet type way, because the relative movement occurs in two directions: perpendicular as well as parallel to the crack surface. Loose particles may also get caught between the two faces of the crack, preventing it from closing completely when the wall warms up and expands.

Similarly, massive walls of concrete or masonry develop very pronounced temperature gradients across the thickness of the wall, with temperature distributions that may be highly non-linear. Typically, a wall will warm up on its outer surface during the day sometimes followed by rapid cooling due to wind and rain. These surface temperatures will typically taper out at a distance of 40 to 60 mm under the surface. During the following few days, or when the exterior face cools during the night, some of the heat the concrete has received will

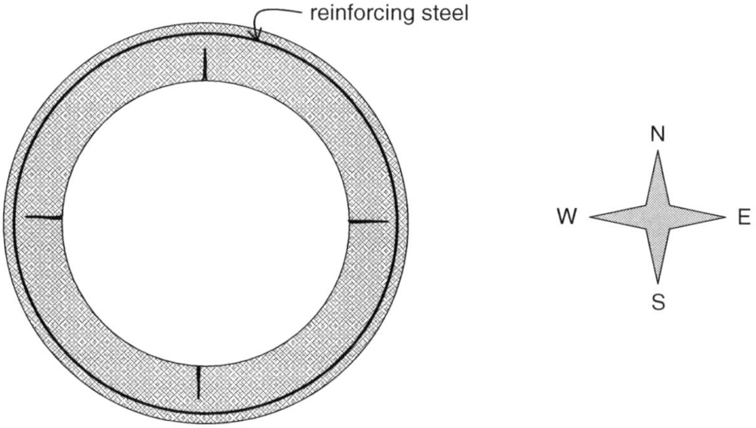

Fig. 9.2: Temperature cracks in a cylindrical tower wall

migrate to the interior as a kind of temperature wave, slowly dying away at a depth of 150 to 200 mm or so. Measurements during a summer have shown that the daily and even shorter variations were "seen" by a thermometer 25 mm inside the exterior surface, but that near the inside face, only weekly or longer-term trends showed up. Those measurements were taken on a cylindrical concrete tower (Moncton, Canada) with walls 12 inches (305 mm) in thickness. There was little communication of the inside body of air with the outside, and no noticeable movement within the same.

Four vertical cracks had developed in the tower walls over time and were oriented almost precisely in the wind rose direction (see *Figure 9.2*). Because no horizontal reinforcing had been provided near the inside surface, they had progressed to a state where pieces of concrete started to spall off the faces of the cracks and fall.

Similar cracking patterns have been seen on a number of masonry structures with massive walls (0.5–1.5 m) forming a closed cylinder or a square.

At first, this condition is not very serious, structurally, as long as the cracks do not open beyond a certain width where a minimum mechanical interlock still exists among the separate portions of the structure. Beyond this, however, the structure is no longer *integer* but is *coming apart*; it may no longer be able to resist horizontal loads due to wind and, above all, earthquakes. Falling pieces of concrete may cause a nuisance or danger, and water entry is invited.

In countries with freezing temperatures, water which has entered the wall will freeze periodically, accelerating the process of degradation. If maintenance is not provided frequently and thoroughly, the building may soon be a ruin, as witnessed by the numerous former castles throughout central and northern Europe which are presently in various states of decay or have disappeared altogether.

Modern construction has – rather recently – recognized the fact of these temperature gradients and most builders will presently provide relief to exposed elements of concrete or masonry through joints where relative movements can occur without causing damage or degradation: Robustness is created in this case by removing continuity. Let us note, however, that complete separation can only be provided where no forces must be transferred from one to another of the separated elements. Primary structural systems such as bridges or building structures and elements, which must provide force paths over a certain distance, cannot be allowed to deform freely and individually, and thermal stresses as well as cracking are often inevitable.

In this case, crack control must be provided, meaning that the cracks that will form inevitably must be prevented from opening beyond a certain width, and from widening progressively. The classic way to do this in concrete construction is to provide "temperature reinforcing" near the surface and at close spacing. This has turned out to be of ambiguous value, at least when using carbon steel bars: The closer to the surface the steel is placed, the more effective it will control the width of cracks but the quicker it will be reached by the corrosive effect of water, especially when it contains salt.

The precise position, spacing and diameter of "temperature steel" have largely remained a matter of faith, experience varying from one type of environment and exposure to another. Modern styles of reinforcing using non-corrosive materials are preferable to carbon steel but their cost is often prohibitively high.

Various theories involving the tensile strength of concrete among a host of other parameters describing hypothetically the conditions the building will experience have been proposed to determine the best way to design temperature reinforcing but they remain of mostly academic interest. On the other hand, some owners of large amounts of infrastructure (states, cities) have created their own rulebooks, based on practical experience gathered in the last few decades. A recent trend has been seen, not to put any reinforcement near the exposed surface, for instance in road surfaces made of concrete, but to minimize tensile stresses from restraint, and to optimize tensile resistance in the concrete.

For masonry construction, circumstances are considerably different: Tensile strength is practically non-existent across mortar joints, but stresses due to restrained thermal deformations are much lower. Where deformations caused by thermal scale changes in the masonry (brick, stone, concrete blocks, artificial stone, etc.) are restrained overall, they are mainly absorbed by the mortar joints which are able to do this to a degree and for some time, depending on the type of mortar and the severity of the case. However, all mortars are essentially brittle materials and will eventually crack and disintegrate, leaving the wall open for water entry. The physical description of this is quite simple and easy to understand qualitatively, including capillary action in the porous mortar and ice formation, etc. that may accompany the process. Quantitatively however, much remains to be learned. Differences of climate, even local microclimate, may decide whether a wall suffers damage quickly or only in the long run. Studies on existing buildings have revealed that portions of one and the same wall have deteriorated in years while other parts in seemingly identical conditions remain in perfect condition after decades.

A few general observations can be stated, however, applying chiefly to structures relying on cementitious materials for their coherence, i.e. masonry and concrete:

- Thermal deformations due to exposure to climatic conditions are cyclic and may become partially irreversible and consequently, progressive. Providing expansion and contraction joints at sufficiently small intervals may avoid this.
- Brittle materials such as concrete or masonry will develop cracks where tensile stresses occur. These cracks, if allowed to widen beyond a certain width and depth, will invite water entry, with all the consequences this may have in the particular circumstances (freeze-thaw cycles, etc.).

- Thermal stresses and deformations are cumulative with other, similar effects such as shrinkage or creep. If deformations are accommodated and not restrained, the stresses caused by temperature variation will disappear. Providing freedom of movement may therefore enhance robustness with respect to long-term performance.

9.6 The collision strut (strength in solidarity)

Traditionally, railways have built their installations in or near the centre of cities, including large expanses of freight tracks. Land prices in these locations provide a strong incentive for the tracks to be overbuilt which creates the problem of relatively slender columns being constructed within reach of a locomotive or train leaving its track accidentally.

Due to confined space conditions, especially in areas with switches, making the columns more massive is often not an option so that other ways must be found to create a situation of robustness. The problem was recognized in the past when railroad tracks went underground, and especially in North America, the collision strut was introduced, linking a number of columns in the longitudinal direction, usually in combination with some diagonal bracing in the plane of the columns. The idea is that the columns are solidarized in this manner so that no single one can be knocked out since the shock will be resisted by all of them together. The system does not help much for a push perpendicular to the direction of the tracks but that component of the possible collision force is only a fraction of the longitudinal one, and the columns may be equipped in a different way to resist it.

Where platforms exist between the tracks, i.e. in a passenger terminal, they are mostly built in elevated position (between 0.5 m and 1 m above the tracks) and provide therefore a dual protection, as a collision strut (although not in ideal position which is somewhat higher) on one hand, and as a protective element on the other, since the columns are spatially removed from the oncoming train by half the width of the platform.

9.7 Façades and curtain walls

The structural problems and failures of façades are intimately related to aspects of building physics such as temperature, heat flow, thermal expansion and contraction, long-term deformations due to shrinkage and creep, moisture from infiltration or condensation and its effect on corrosion as well as ice formation. In this brief review, the aspects of structural resistance will be addressed, relating to the functioning of the details attaching the façade elements to the main structure and to each other. Other types of failures, relating mainly to incompatible deformations, are not addressed here.

Façade design has long been a stepchild to the building design community, each party nursing the impression that someone else should/would take care of it. While traditional styles of building envelopes such as massive brick or stone walls were being abandoned for reasons of economics and – more recently – considerations of energy conservation, new styles and concepts were developed and found widespread use, mostly following the Second World War. Where traditionally the building envelope doubled as a structural system, carrying vertical and horizontal loads, the physical separation between the structure and the building envelope is almost complete for modern systems, with curtain wall elements meant to resist only very local loading effects such as wind load onto, and weight and seismic effects of the element itself. These loads are then collected and transmitted to the building structure via small, mostly metallic elements, the detailing of which has received little attention by schools, science and professional organizations so far.

Many of the concepts for curtain wall support are now a number of decades old, and experience with their performance begins to include long-term effects such as corrosion and the consequences of incompatible restraint. The general tendency is in the direction of degradation and – since most of the attachment details are hidden from view and difficult to lay bare – the situation begins to acquire the characteristics of a minefield, with accidents waiting to happen anytime, in particular in the context of older buildings, which were built in an environment of innovation and with little understanding of the causes and effects of building physics. As the large quantity of construction from 30 to 60 years ago is ageing, one may fear that the frequency of failures will increase rather than decline, triggered by such effects as wind storms, minor earthquakes or exceptionally low or high temperatures.

In the case of older buildings, structural intervention usually comes up when major renovation work is undertaken; typically architecture and mechanical/electrical systems are replaced or renovated every 30 years or so, often leaving the structural system mostly intact. This then becomes the opportunity to view, test, verify and correct, modify or replace curtain wall attachments and supports. The existing elements can now tell their story and the architect or engineer may learn what has worked and what has not, and why. Where failures have occurred, renovation work will be triggered and corrective measures become the only option, regardless of narrow-minded economic considerations.

Most curtain wall failures related to the attachments are of two types:

- corrosion
- anchorage failure

Both involve a weakening of the resistance of the particular attachment detail and are progressive in nature. Often they are found to combine where corrosion products burst the brick, stone or concrete around the anchorage. Rust is typically five to ten times as thick as the steel it was, and develops considerable pressure when it forms, which usually exceeds the tensile resistance of brittle materials such as concrete.

As for the quest for robustness, we are now in a situation of hidden flaws, which develop and grow over time, and it would seem to be a good idea to try and find a strategy to establish the comfort one would like to experience with a building design.

Our new designs of curtain wall attachments will of course be directed towards correcting the flaws which came to our attention but essentially, they will be in the same situation as before: hidden from view and exposed to the effects of variable temperature and humidity the precise reality of which is hard to predict. Only time will tell if our concepts were good or not so good.

Robustness being the property of systems making them stand up to any, including the unpredictable, effects, the hypothetically critical scenario in the case of curtain wall attachments would then be the degradation of the resistance of one or more of them to a point where it (they) can no longer resist the forces flowing through them. Among the possible strategies to provide robustness, the following approaches come to mind:

- Try to make the back of the curtain wall inspectable, if only in a sample fashion.

- Do the utmost possible to protect the metallic parts against corrosion. They must function in an extremely unfavourable environment (variable mostly high humidity, ample supply of oxygen, very large temperature variations). Stainless steel may not be the right answer here; bad experiences exist with commercial quality *stainless steel* when left in humid places and under permanent stress (see also Ref. [16]). If details include welding, protective coats should include cathodic protection through a sacrificial anode (Zn, Al). If galvanized parts are used, no field welding should be allowed.
- Structural measures are more difficult to specify but may be more effective to prevent accidents (pieces falling off or other types of failure). As a general approach redundancy combined with deformability seems to provide the best answer. Curtain walls that show their distress in terms of dislocated elements or so, give warning and can be repaired in time; this requires details which permit movement before loosing (all) resistance. Deformability can be achieved not only through ductility, for example bending of plates, but slipping bolts in oversize holes may do the same job as well. The important thing to note is that deformability cannot usually be found in the material of the façade itself, which is made from brittle materials (masonry, precast concrete, glass, etc.) nor can it be provided by the main structure supporting it which is of a calibre making it practically rigid for the small forces in the building envelope. Therefore, true to the fuse principle of capacity design, the resistance to pull out or cracking at anchorages must be greater than the force which makes the attachment detail yield or move at bolted interfaces.

If the critical scenario includes the loss of resistance in one or several connections, the remaining ones must be sufficiently strong to resist the forces acting, and sufficiently deformable to permit the redistribution of forces. In certain cases, the curtain wall itself may help to bridge over missing attachments, through arching action: Consider a masonry veneer supported by a shelf angle which in turn is attached to the structure behind at a certain spacing, say 1.5 m (see *Figure 9.3*). Let us assume that one of the attachments has disappeared in terms of resistance (through corrosion for instance). The shelf angle must now support the weight of brick over a 3-m span which, if that weight acts uniformly, will make it deflect, perhaps to the point of plastification or torsional instability. It will however not happen like this in all probability because the brick will not follow the beginning deflection of the shelf angle, but will form an arch-type compression field bridging over the lost attachment. It may crack however, indicating that a problem exists.

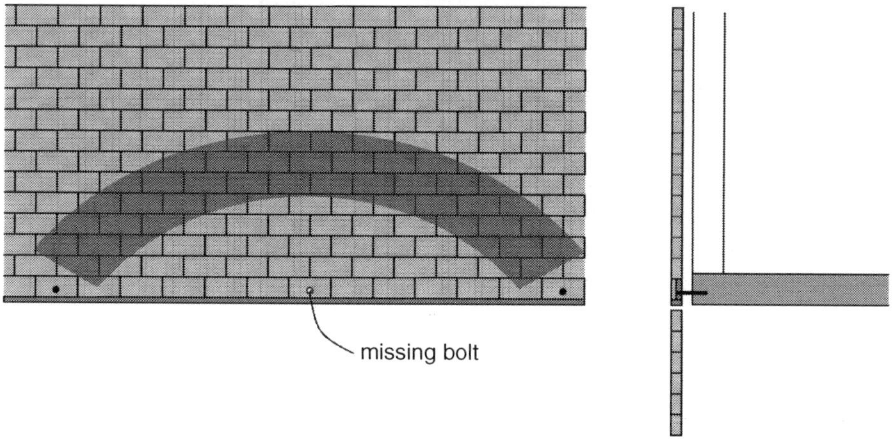

Fig. 9.3: Compression arch forming over missing point of resistance

Similar things have been seen at corners where a downward movement of the shelf angle produced a horizontal tension field at the top of the masonry because it is in a cantilever situation. This is not something engineers are comfortable with, but another feature comes to our aid: If the tension resistance fails, the cantilevering part of the veneer wall will settle onto the veneer of the next storey below, the respective anchorage of which will now have to carry twice the weight if robustness is to be maintained. It is therefore a good idea to design the anchorages near corners for an overstrength equivalent to, for example, a load factor of 2.5 to 3 or so.

9.8 Earthquakes and unreinforced masonry (second line of defence)

A great quantity of buildings with bearing walls of unreinforced masonry exist all over the world, many among them of heritage value, and many of them located in regions of strong seismic activity.

Unreinforced masonry possesses properties which vary greatly and which are not in good agreement with modern ideas about seismic resistance although methods do exist and have been employed in some countries to make masonry walls more robust. However, there are limits to this and most masonry is of a rather brittle character where, beyond a certain rate of shear deformation, resistance is lost rather quickly through progressive disintegration of the walls. In order to improve on this state of affairs, or in other words, to provide robustness during a strong earthquake, it is most important to recognize as the key criterion of masonry behaviour that deformations must be limited effectively.

Tolerance to shear deformation γ varies considerably depending on the material and style of the masonry but under cyclic loads it is limited to values of $1\% < \gamma < 3\%$ or so, for common styles of masonry. Translated into stiffness requirements, this turns out to be a rather stringent criterion and the designer must carefully and quantitatively validate any options he may wish to consider, establishing that they will really work. This may include recognition of the particular character of the earthquake to be expected, i.e. the frequency and displacement content.

The practice knows a number of methods to enhance the seismic performance of masonry structures:

- addition of steel bracing
- addition of concrete on the face of the wall
- linking with other structural elements
- base isolation
- enclosure in concrete or steel framing.

Each one of them has its own potential problems and caveats, which must be heeded so that the system works, but a number of rules can be spelled out which are common to all of them albeit in different ways:

- The masonry must be in good repair. If it is not it may disintegrate prematurely, i.e. at lower rates of deformation than anticipated.
- In order to provide real robustness, the composite structure must possess sufficient stiffness to limit deformations to what the masonry can tolerate. This sounds quite straightforward,

even trivial, but in reality it requires knowledge about the behaviour of each structural element participating and this well into the inelastic range. Consider the masonry itself. It will act more or less elastically up to a certain load/deformation where its maximum resistance is mobilized. Beyond this point, its resistance will decay, slowly at first and progressively thereafter. Subsequent loading cycles will change the initial behaviour, with the decay of stiffness setting in earlier, i.e. at lower rates of deformation. It will be entirely different for the reinforcing element. If it consists of a steel brace mounted on the surface of the wall, it will have a stiffness that is a fraction of the initial stiffness of the wall, so most of the load will find its way into the latter until its resistance becomes exhausted. This is where the critical stage is reached: If the reinforcing element does not have sufficient stiffness and strength to control the deformation of the weakened composite system, robustness does not exist.

- The interaction of two or more elements depends on the internal coherence of the system so that the variable forces exchanged among the elements can travel without causing additional distress, i.e. failure at connections, in diaphragms, etc. It is a good idea to *overdesign* these link elements, i.e. to generously provide strength – it may be very difficult to estimate forces acting at each stage and a structure where all elements do not work as a unit may be worse than the unreinforced original.

Reinforcement of an essentially brittle system with a more ductile supplement, in the sense of a second line of defence may not be ideal, i.e. not as good as a newly conceived structure designed for ductility. However in cases where the conservation of existing elements imposes restrictions, it may be the next best thing.

Some other aspects must be considered in a practical design:

- Buckling of the steel members must be prevented. Sometimes the wall itself can serve as the lateral support if it is supported sufficiently by floor plates or cross walls.
- The forces acting in the diagonals must be provided with a force path in and out of the steel members, through anchorage to the rest of the structure.
- Connections must be stronger than the yield strength of the steel members to provide ductility.

The fact that masonry walls usually possess, due to their volume, a very high initial stiffness means that in any combination with other resisting elements, they are a prominent participant in the sharing of lateral forces. Because their share will drop rapidly when deformations exceed a certain limit, the remainder of the system should be able to effectively pick-up the contribution of the masonry so that deflections and excursions remain controlled.

9.9 Assembly of steel structures

The absence of robustness in certain steel structures has been at the origin of numerous accidents in the past and continues to do so. The properties and behaviour of fabricated steel members being quite well known, it is mainly the connections which turn out to be the critical element, and it is fair to say that considerations of robustness should concentrate here.

Unlike timber construction where relatively brittle elements must be assembled with ductile attachments, steel members are usually ductile when used in tension, bending or even compression, if buckling can be excluded. The connections however, including the zones immediately adjacent, often do not possess this property. Bolted connections having been examined in Section 6.11; in the following text mainly welded assemblies will be discussed.

The essence of design for robustness is to endow the final product with that property, i.e. in reality and beyond all theory, standards and specifications. It has been recognized, following some bitter experience, that in the case of welded connections, particularly, a dramatic difference may exist between the models of the future structure as it appears in calculations, drawings and specifications on one hand, and the real life products on the other. The reason for this is that welding is a very complex trade, or it might fairly be called an art, requiring artists for the execution. To confirm this, one has to look only at the volume of literature, codes, standards, work specifications that have been produced on welding, and at the very stringent and highly organized examination procedures for welders. All of this can be understood as an attempt to compensate with quality control for the effect of *Murphy's Law*, which in the case of welding, is of particular importance. In some cases, quality assurance protocols have become so tedious and complicated as to become counterproductive and unreliable themselves, resulting in large amounts of paper, providing "satisfiction" rather than real "bite", leaving hidden flaws.

There is another answer to the problem than complex and onerous quality control:

- Design the assembly so that welds and especially field welds are not located in critical places.
- Use welds that are easy to execute, in terms of access, size and volume.
- Try to avoid welds that require much precision and preparation – under time pressure this may not be possible to achieve.
- Try to protect welds from overloading by providing deformability nearby.
- Use bolted connections where welds would be difficult, and difficult to control. The necessity of field welding should always be carefully examined. Its advantages must outweigh the negative aspects related to reliability and uncertainty.

All of this can be summed up in what amounts to an old engineers' maxim: Try to design around critical situations that scare you and require extensive protocols.

A few examples shall illustrate this:

Field welds are difficult to control effectively, especially in a competitive environment. It may be a good idea to avoid problems through design rather than control mechanisms (see e.g. *Figure 9.4*).

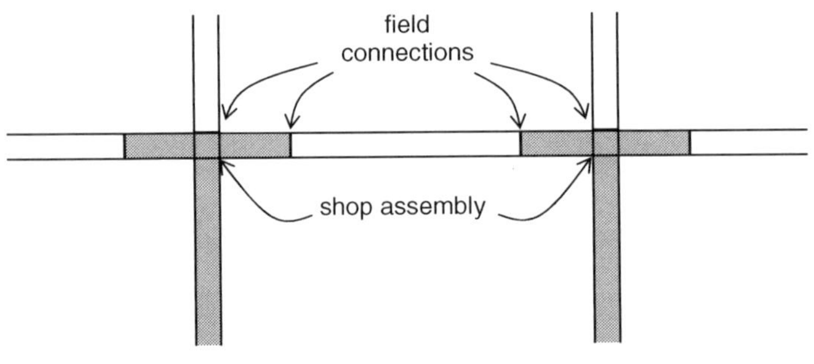

Fig. 9.4: Field welds away from critical places

Some of the following examples relate to an environment where high quality and strict quality control can not be had, i.e. low and medium cost construction for difficult working conditions. The authors are aware that in certain circumstances ideal or nearly ideal conditions can be brought about but usually the direct or indirect costs of this makes it the exception rather than the rule.

Often beams must be assembled in pieces and welded in place. To provide a design that is simple and easy to execute will contribute to safety and robustness. Fillet welds can be proportioned to provide extra strength in order to compensate for less than perfect execution (*Figure 9.5*).

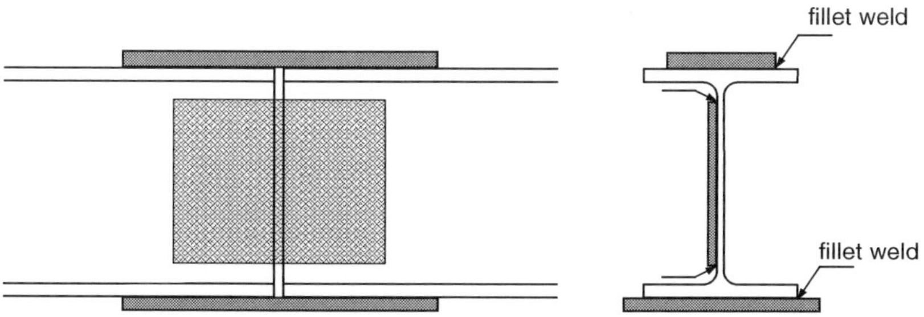

Fig. 9.5: Welded beam splice

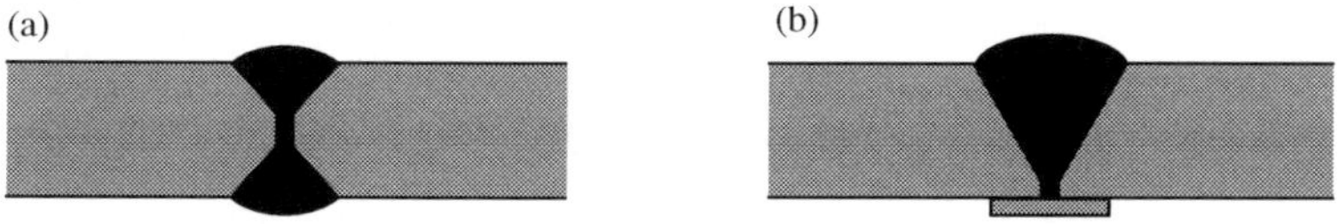

Fig. 9.6: Butt welding of thick plates (a) welded from both sides, minimizing weld deposit, (b) welded from one side only

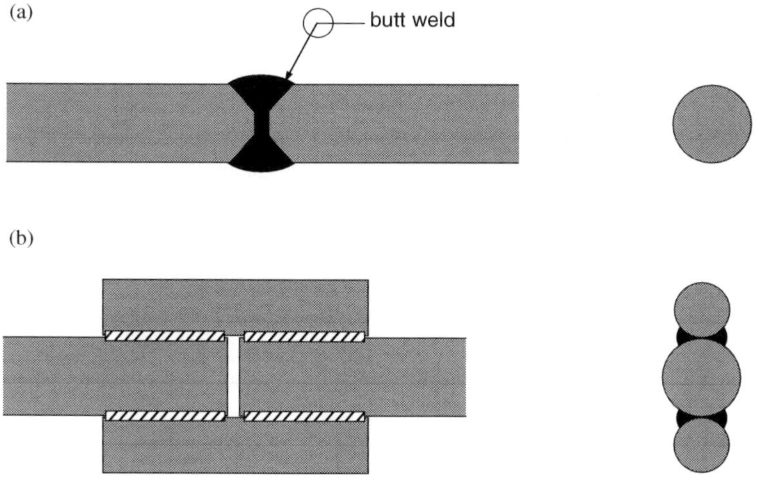

Fig. 9.7: Splices of reinforcing steel (a) Butt weld, (b) Fillet welds

Problems related to welding are a matter of dosage. The bigger the weld the more it will be prone to defects, residual stresses, etc. Minimizing a weld implies to turn the work piece to avoid overhead welding (*Figure 9.6*).

For reinforcement bars butt welds are difficult to execute in many situations since the welding rod must go around the bar. Incomplete welds may compromise the effective strength (*Figure 9.7(a)*). The fillet weld alternative (*Figure 9.7(b)*) is easier and again, welds can be made longer, to suit the scenario.

Tubes assembled directly as shown in *Figure 9.8(a)* must be cut with great precision and the strength of the tubular members cannot be matched in the connection. The gusset plate version (*Figure 9.8(b)*) permits more freedom for the design, and the connection is no longer the critical element.

Providing knee braces instead of complicated welded or bolted assemblies is often an ideal way to create an effective moment-resisting frame, in particular, in the context of upgrading

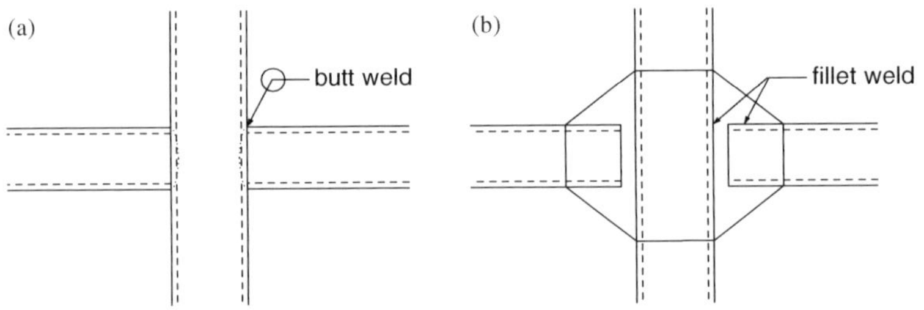

Fig. 9.8: Tubular assemblies (a) without gusset plate, (b) with gusset plate in slots

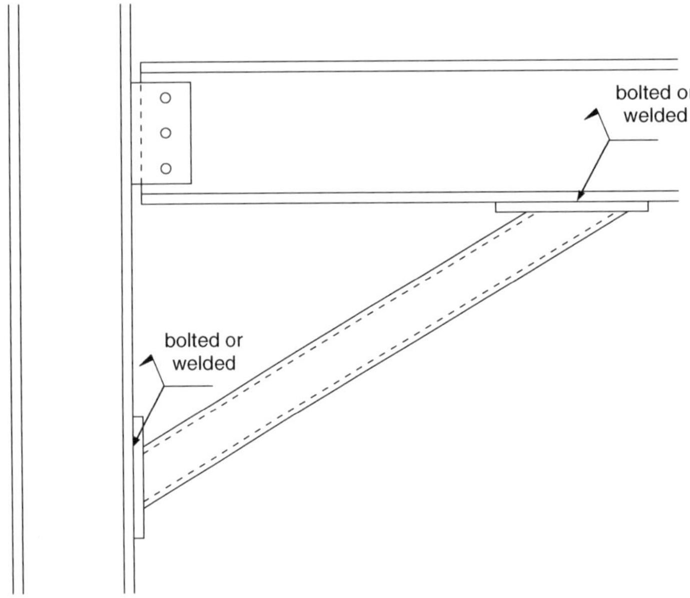

Fig. 9.9: Moment-resisting connection with knee brace

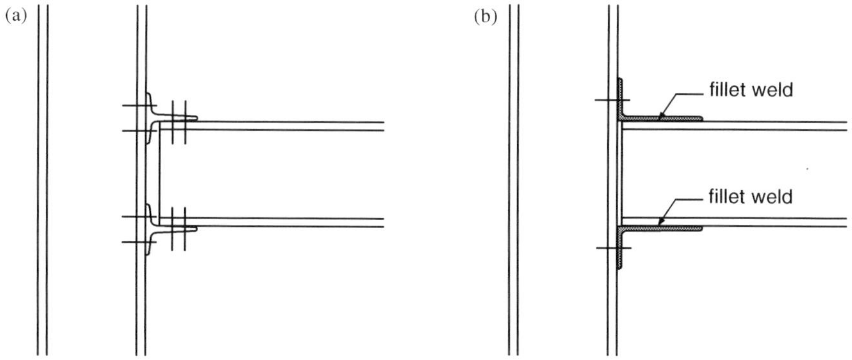

Fig. 9.10: Two examples for ductile moment-resisting connections (a) riveted detail found in older building frames, (b) bolted and welded

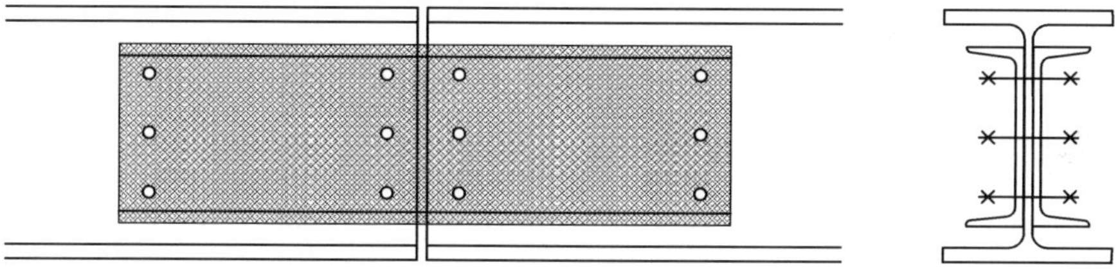

Fig. 9.11: Bolted field splice

existing structures. Ductility can be designed as required (*Figure 9.9*). Of course, the existing beam column connection must be able to resist the forces generated in the modified scheme. Additional reinforcement in the floor may be needed.

Often it is not practical to make a connection stronger than the connected members themselves. To provide deformability in the plastic range, bending of plates or angles can be used (*Figure 9.10*). Nevertheless, the connecting angles and T-beams must be sufficiently rigid to avoid excessive prying of the bolts.

The splice detail shown in *Figure 9.11* is ideal for use in existing buildings where materials must be entered by the service elevator. Maximum length is usually limited to approximately 4 m. Its execution does not demand any special skill and robustness comes without any special provisions.

Summarizing the essence of *designing around problems*:

Try to avoid critical welds requiring extensive and complex quality control, or special skill, because you may not get it in real life. In particular, avoid butt welds in places of maximum tensile stresses. Butt welds cannot be oversized for robustness (fillets welds can, in the direction of the main stress).

Very often a small amount of additional material can be introduced to avoid problems with safety, reliability and robustness. This applies in particular to situations of custom made *one of a kind* connections where simplicity and ease of execution are directly related to the robustness.

9.10 Space trusses on point supports (multiple load paths and their problems)

space trusses, or *space frames,* as architects call them lovingly, possess some inherent properties, which apply quite typically, even generically to all systems, proprietary or custom made as they may be:

- Slender members, with l/r typically somewhere between 100 and 200 (l = buckling length, r = radius of gyration).
- Connections the resistance of which, especially in tension, is considerably weaker than the resistance of the member which is connected; various types of connections are offered by the market, with through bolts, sockets, cut threads, welds, etc., most having in common that each member has a reduced cross section at its end in order to fit to the joint assembly the overall size of which is critical for aesthetics, and must also lend itself to quick and easy installation.

For both compression and tension then, one has a structural system made up of essentially brittle elements, making robustness difficult to achieve.

To aggravate the situation, space trusses are often erected on point supports, in order to achieve an aspect of openness with the roof floating overhead, and the space underneath open to view and air circulation in every direction. This creates a condition rather similar to the situation which was discussed in the context of flat plates, with a concentration of forces around the head of the columns, the only difference being that in this case, those forces exist in the form of axial loads in the primary diagonals attached to the column head where they were modelled as shear stresses in the concrete.

Aesthetics again tends to dictate that the size (diameter) of these primary diagonals be the same as for the rest of the structure, i.e. as slender (elegant) as possible. One way to achieve this is to choose a thicker wall pipe or a solid cross section with the same outside diameter. The degree of slenderness is increased hereby, making these diagonals effectively more brittle in the case of compression. In the case of tension (see *Figure 9.12*), the connection will always be the weakest link.

Two-way space trusses supported by an array of columns are highly hyperstatic systems and the distribution of loads is influenced by imposed deformations such as unequal foundation settlement, and by the distribution of weight on the roof. This condition is most important in countries where the weight of the snow may represent a large proportion of the total load, and is subject to accumulations, the size and location of which are difficult to predict. The situation may be further aggravated for the diagonals joined to the column head were the assembly acts as a *sway frame*, resisting horizontal loads. Considerable uncertainty therefore exists about the real share of the load, which must be resisted by each of the diagonals at the column head. One can construe numerical examples using normal case descriptions, which demonstrate this to be quite dramatic, one particular member being called to resist most of the column load by itself at one time or another. Qualitatively, thus a number of circumstances are commonly coming together to make space truss systems particularly vulnerable, in particular, concerning the primary diagonal members which are guiding the load to the column.

The question is then, how robustness could be achieved in spite of the properties of the structure that tend towards a brittle type of response, on several counts?

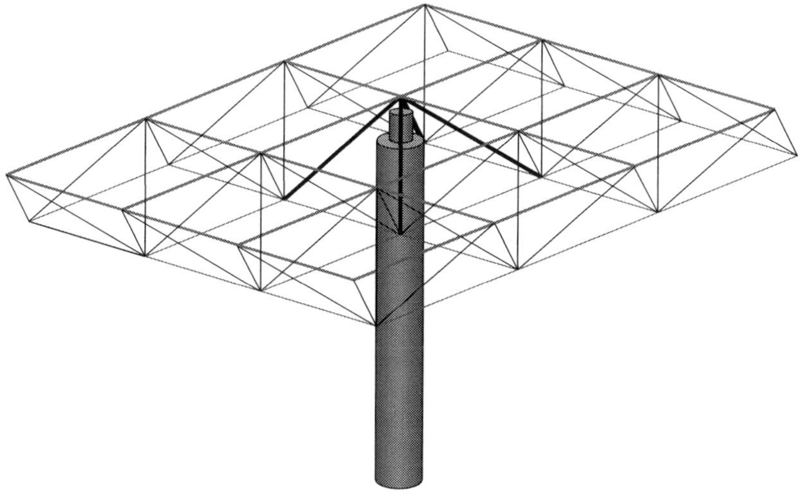

Fig. 9.12: Reversal of main diagonals for better robustness

The approach of choice would obviously be to provide ductility to the primary members, i.e. the diagonals in question, so that they will work as multiple load paths, effectively. Since this is only possible if they are in a tension situation, they must be oriented so that they meet the column at the top rather than at the low point (see *Figure 9.12*).

The end connections must also be detailed for higher strength than the shaft of the diagonals that must yield first since this is the only place where plastification on a sufficient length can be mobilized. Not so at the connections: Even if they are able to plastify, i.e. sustain a certain level of resistance through a range of (permanent as it turns out) deformation, the absolute value of this extension is severely limited as it takes place in a very confined volume of material, e.g. yielding of a short length bolt, at a weld or in a piece with reduced cross section. In order to permit a substantial redistribution of load from one diagonal (the one becoming overstressed first) to its neighbours, it must be capable of a finite plastic stretching (i.e. waiting for its companions to pick up the excess load).

Details have been developed to provoke first yielding in bars, away from the connection in different contexts, e.g. mechanical splices of reinforcing bars, with rolled-on threads, or providing thickened ends where threads must be cut. Similar assemblies can be thought of, exploiting the strain-hardening properties of good-quality steel, where *good quality* in this context means that a good plastic extension (e.g. 15% on a length of 200 mm or so) is followed by a substantial increase in tensile resistance (i.e. strain hardening, for example 20–25%).

What may be practical for reinforcing bars is not necessarily applicable to a space truss assembly, and most examples of this type of structural system seen today are still of the essentially brittle type, with little or no robustness.

An alternate avenue to provide more robustness to space trusses is obviously to increase the strength of the critical members, e.g. based on an analysis of the most unfavourable loading scenario. This will result in an increase in the size of the critical members, which in some cases may mean all members, in order to provide uniform appearance and assembly details. It may also be contrary to architectural criteria of elegance.

An increase in strength is equivalent to an increase in safety margins, and in situations of multiple load paths with non-ductile elements, the reserve strength implied by higher safety margins may be the only effective way to provide resistance in unforeseen circumstances, i.e. robustness.

A simple calculation, similar to the consideration in the *hanger case* (Section 9.11) may demonstrate this:

Normally, steel elements are dimensioned to an effective safety factor of 1.5 to 2, including factoring of loads and material strength. The loss of one member out of three due to overload will, in an idealized situation, transfer the load P_{trans} from the failed element to each of the other two.

$$P_{\text{trans}} = \frac{1}{2}\gamma P_0 \text{with} \gamma = \frac{R}{P_0} \tag{9.1}$$

where P_0 is the unfactored design load for each element, γ the safety factor and R the ultimate resistance. With a safety factor $\gamma = 1.5$, the remaining two elements will receive

$$P = P_0 + P_{\text{trans}} = P_0\left(1 + \frac{\gamma}{2}\right) = 1.75\, P_0 \tag{9.2}$$

which may exceed their ultimate resistance. To provide sufficient resistance to the system, a higher safety factor is therefore needed. In the theoretical example at hand, this turns out to be $\gamma = 2.0$ as a minimum.

This does not account for a number of circumstances, which may accompany the event:

- More than one member may receive loads higher than the design load P_0.
- Shock effects will likely accompany the sudden rupture/buckling of a primary element.
- The remaining elements may be in a position within the structural system, where they receive more than just their share of the ultimate load of the failed member. The modified structural system without the failed element will have a different pattern of internal forces, possibly putting the remaining elements at a relative disadvantage.

In the case of the flaw event, the situation may be somewhat less serious because the flawed element will fail at a lower load than its theoretical ultimate resistance, due to the flaw. Since the quantitative effect of the flaw is usually unknown, it may be prudent, however, to consider the case in the same light as the event of overload.

Morphologically the case of the critical diagonals of the space truss is similar to the situation of the wood joists: Multiple load paths and (relatively) brittle elements. The essential difference here is quantitative and due to the large scatter one finds in the mechanical properties of wood, especially sawn lumber which is chiefly used for joist floor construction. This has led to considerably higher safety margins, in terms of strength reduction factors than for steel, which is paramount to a greater strength reserve.

$$\text{Strength reserve} = \text{strength} - \text{design load} = (\text{design load}) \cdot (\gamma - 1) \tag{9.3}$$

Using nominal values for strength and load, one finds for a normal steel construction $\gamma - 1 \approx 60\%$ and for reinforced concrete $70\% < (\gamma - 1) < 100\%$, depending on the composition of internal forces (bending, compression shear, etc.), and $(\gamma - 1) \approx 200\%$ or so for wood.

Wood in bending is also considerably less brittle than steel members of high slenderness, or with weak end connections.

9.11 Hanger elements (multiple load path)

In many circumstances, parts of a structure or other weights are supported by hanger elements such as cables, rods, chains, etc. Hanger elements are usually small in cross section as there is no question of buckling, and made from metals with good ductility properties.

They are also very vulnerable to accidents such as fatigue, shock, corrosion, (which in some metals is enhanced by permanent tensile stress), to wear and tear and to destructive action, be it intentional or not, and not to forget to flaws of any kind.

In the same class one could count fasteners acting in tension, the cables and hangers of suspension bridges, the stays of a cable stayed structure, and eventually, tension diagonals in trusses.

All these applications have in common that the tension member is situated on a major if not the only force path and its rupture has very serious consequences, at least in all force controlled conditions involving gravity loads, or any loads which are invariant with regard to the response of the structure.

An old rule applying to this situation is: "one nail is no nail", or for older engineers who were educated to riveted steel assembly: "one rivet is no rivet", later modified to "one bolt is no bolt". This is all nice and dandy and no steel handbook will show connections relying on one bolt only.

The problem is that similar situations can come about in many guises and major accidents have been associated with this type of scenario.

Consider the architect who wants the lobby of his building column free, creating an effect of grandeur. The obedient engineer will then be forced to suspend balconies, partial floors, etc. from above which creates the hanger scenario. And here is where the horror story begins: The hanger member being out of character with the rest of the architecture, it will likely be hidden in a wall or otherwise disguised. Enter the electrician who has to pass some of his "spaghettis" through the wall. He will need openings to be created when the structure has been completed some time ago and may already be hidden from view.

It is easy to think up equally horrific scenarios involving such things as a high-strength steel member touched by weld beads or heated beyond the point of loss of strength from cold drawing, or being chafed by something rubbing against it. Secondary bending moments may – unnoticed under repeated loads much inferior to the design loads – cause fatigue which in fact amounts to a loss of strength as well as ductility, followed by a sudden rupture. Corrosion may reduce in time the safety margin.

The answer to all of this is quite straightforward: Do not put all the eggs into the same basket, in other words: Provide more than one element, following the common sense rule.

$$\text{Strength}\begin{pmatrix}\text{All parallel}\\ \text{elements} - 1\end{pmatrix} > \text{Minimum capacity required} \qquad (9.4)$$

The minimum capacity may be somewhat lower than the factored limit state stipulated by the building code but should still include some safety margin to cover other eventualities. To provide two or more elements in parallel is not very costly since little material is usually involved. Also, if more than two elements are provided, the additional – expendable – element represents only a small differential. Where this approach does not work, the next best thing to consider would probably be to provide extra strength, or size so that a small defect such as the electrician's hole will not be fatal.

On truss bridges, an arch-shaped top or bottom chord will greatly reduce the importance of the diagonals and the bridge might survive the loss of one of them.

9.12 Structural glass

Glass is a typical brittle material. Subjected to short-term tension or bending, it shows a linear elastic behaviour until cracks occur. Cracks can be initiated in different ways and due to various causes; and they propagate rapidly wherever tensile stress is present and may – depending on the structural system – lead to the failure of the structural element.

Glass has always been a building material. Until recently, however, its structural applications have been quite limited. Vertical panes as they are used for windows carry the self weight as in-plane compression. Only lateral forces such as wind effects have to be resisted in flexure. The panes are usually supported individually so that they do not have to carry loads from above in compression. Glass panes used for roofs of moderate inclination have to be designed for a snow load, which again means bending of the panes. As long as the dimensions of the panes remain within certain limits, the consequences of a failure of a pane are relatively modest and can usually be accepted.

Modern architecture has moved to extend the use of glass in several directions [2, 17, 19, 21]:

- Window panes have become larger, up to the limits imposed by production processes.
- Façades are becoming higher. The failure of even a small pane can have serious consequences.
- Glass roofs have to be accessible for cleaning and maintenance, i.e. they have to be able to support single loads of about 1 kN.
- Glass floors are made accessible for pedestrian or even vehicular traffic, permitting illumination or even views from below.
- Glass is used for stairs, walls, beams and columns, i.e. as structural elements having a local or even a global bearing function within a building.

Those latter applications are summarized as structural glass [7], and similar or equivalent requirements apply as for structural elements with other materials such as steel or concrete.

To provide robustness, several basic avenues exist:

- To provide strength, through increased thickness, or better quality. The statistical strength of glass possesses a larger scatter than most materials, with a few "bad" pieces having to be considered acceptable in many situations. Quality control measures can reduce the frequency and risk of breakage.
- In tempered glass the zone of tensile stress is removed from the surface where most cracks are initiated through imperfection or wear and tear, and transferred to the interior through internal stress condition produced by the heating and cooling process. This results in increased strength.

- Laminated glass has proved to be a suitable choice for many applications of high exposure such as windshields of cars, glass floors, stairs, etc. Two or more plates interfaced with a softer material are joined together in a composite. It is interesting to note that it is usually the plate most removed from the contact with an impacting agent, which breaks: It is the one which receives the tensile stress. It should also be noted that this does not hold true for permanent or long-term loading where the glass plates are acting individually, no longer as a composite.

The commonly used glass types are float glass, heat-strengthened glass and tempered glass [8, 18].

Float glass is cooled slowly in order to minimize residual and local stresses. With respect to robustness, this means that by loading, the tensile strength at the surface may be reached, but the cracks propagate more slowly, leading to a satisfactory overall behaviour in many applications.

Heat-strengthened glass and tempered glass have been heated and rapidly cooled at the surfaces. This produces tensile stresses in the core when it is cooling in turn, wanting to contract against the resistance of the exterior zone which has cooled and hardened earlier, and is therefore being compressed. Bending stresses caused by out of plane loading are then superimposed onto this "prestressed" state of stress, and surface cracks can occur only when the surface stress becomes tensile. Depending on the heat treatment protocol, the character of the breakage varies considerably, in particular, concerning the typical size of the shards, but is always sudden and without warning.

The robustness and strength of glass elements is considerably increased by laminating several panes to a composite cross section. As an interlayer, sheets of polyvinyl butyral (PVB) are used, which show viscoelastic behaviour [20]. Since the PVB-foil does not only act in shear but also in tension with a large ultimate strain, laminated glass shows a fundamentally different behaviour. *Figure 9.13* shows the three stages of a pane consisting of two layers of equal thickness. As long as both layers are intact the pane is in stage I. For short-term loads shear stiffness is high and for long-term loads it mainly depends on the temperature. As a simplification, plain strains i.e. perfect composite action can be assumed for short-term loads. Composite action for long-term loads on the other hand does not exist for practical purposes. With increasing bending the tension eventually reaches the tensile strength surface and rupture occurs, reaching stage II. Now, the upper layer has to carry the entire bending moment until the tensile strength is also reached at the interface to the PVB-foil. With float glass, the occurring

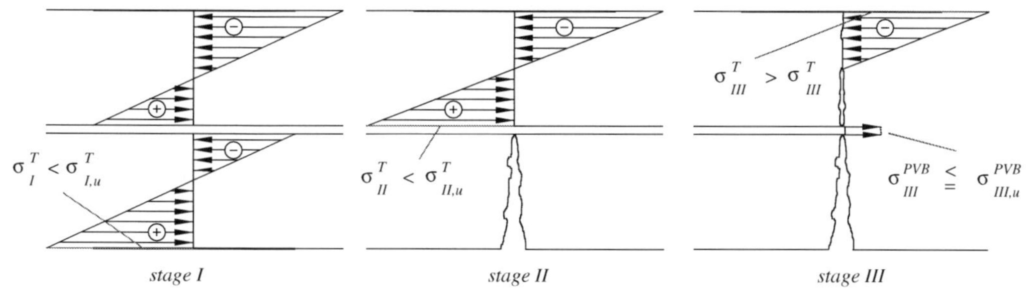

Fig. 9.13: Three stages of behaviour in flexure I, II and III with related stress distribution (according to [9])

(a) (b)

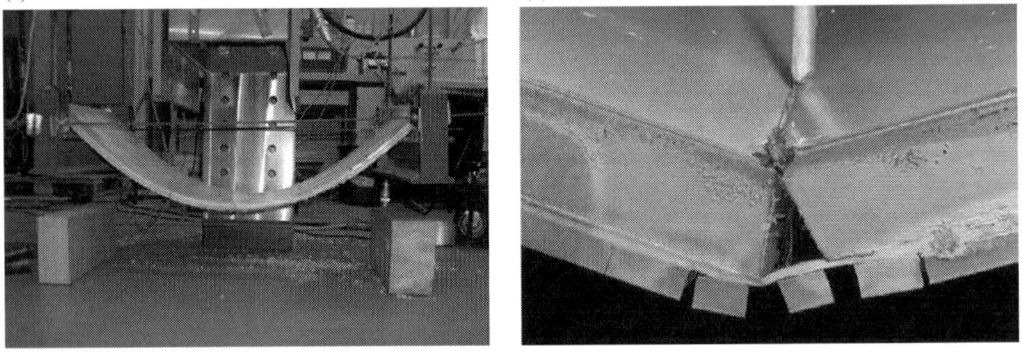

Fig. 9.14: Broken laminated safety glass; (a) reached curvature, (b) concentrated curvature in a yield line (Courtesy to A. Kott)

cracks remain limited to the zones subjected to tension and the upper regions can still act in compression (stage III). The PVB-foil now acts in tension-like reinforcement in concrete. As shown in *Figure 9.14(a)*, a considerable curvature can be reached. Like in concrete slabs, large rotations will be concentrated, forming yield lines (*Figure 9.14(b)*). Further details are given in Ref. [9].

As a conclusion it can be stated that laminated float glass should be used for all applications, where glass has a structural function and where failure or even falling shards could endanger life and limb.

9.13 Fabric structures

Tensile roof structures made from fabric have become very popular following some high-profile applications such as the Munich Olympic Games Stadium and the Haj Terminal of the International Airport at Jeddah both from 1972, etc. As is customary in the trade, every builder wishes to better his predecessor, constructing something more spectacular which quality often translates into *bigger*. It follows logically from this, as with every new technique, that it will be taken to its limits which are learned from the reality of the application, in other words, from one-to-one models or *prototypes*, reality not being known precisely enough to be modelled representatively for digital (computational) or analogue (scale models) analysis.

One of the most impressive examples of this state of affairs is the roof of the Montreal Olympic Stadium which has failed twice and dramatically enough to warrant replacement of the entire roof each time.

The first roof was built with Kevlar fabric (Aramid strands held together by hydrogen bonds). It ripped during a summer thunderstorm, which in terms of recorded wind speeds was not exceptional at all; the experts identified the critical event associated with the failure as a *microburst*, i.e. a local phenomenon embedded in the storm.

The second roof, which replaced the first when it was condemned as too risky, was made from fibreglass fabric, embedded in a matrix. It ripped dramatically following a snowfall with some melt water ponding in a relatively flat portion of the roof. Again, the roof although repaired, had to be condemned and the stadium cannot be used in the snow season.

The notable features of this case, relating to robustness, or its absence as it were, are the following:

- The Montreal stadium roof is the largest of its type constructed to date, by a substantial measure.
- It is located in an area of relatively violent climatic events.
- It was, in both cases, built from fabrics of an extremely brittle nature.
- No attempt had been made to compensate for, or to mitigate the effects of the brittleness of the basic material.

The brittleness of the fabrics in question (other fabrics with different properties exist but could not be used here for other reasons such as creep and lack of durability) is due to a number of reasons, which can be discussed separately, although of course their compound effects are decisive for the performance of the structural system:

- The total absence of ductility of the fibres.
- The tightness of the weave. In a fabric which is loosely woven, the individual threads, although they may be made of brittle material, can stretch by straightening on, rearranging themselves in such a way that the structural system, i.e. the fabric, becomes less brittle than the threads themselves in direct tension.
- The fact that in a fabric structure all parts of the membrane must or should be in tension in two directions prevents the fabric from making folds, which would permit the threads to adjust their length and become solidary through bunching up.
- The absence of *strong lines* such as reinforced bonds, or embedded cables made from stronger and/or more ductile materials (steel, aluminium, etc.), which would prevent a progressive (zipper type) failure.

In terms of material properties for the fabric, two values need to be determined and tested, both being affected also by aging:

- The tensile strength. It must be tested in a carefully balanced way, not to be affected by the more important.
- Tearing resistance (*Weiterreissfestigkeit* in German) for which special testing methods have been standardized (ASTM D4851, D1424, D2261). Typically the tearing strength is given in terms of *force* rather than *force per unit width*. It is representative of the force necessary to propagate a rip, and for brittle-style fabrics it can be very low indeed, so that they can be ripped by hand. Older people may remember the way silk was separated by the salesperson in the fabric store, silk being a strong but very brittle fabric as well.

Fabric structures have therefore a rather common problem with robustness, as was learned in the case of the Montreal Olympic stadium – the fabric style roof is finally being abandoned in favour of a fixed sheet metal style roof, after the failures of the two different fabrics, associated with two different climatic events, both of which were, in terms of recorded intensity, well below the anticipated or codified levels. In both cases, safety factors used in the design to determine the necessary tensile strength had been of the order of 3 to 4, relating to calculated stresses, but failure occurred in a zipper style as is to be expected for multiple load path systems with brittle elements.

Other failures of fabric structures are indicative of a relatively new technology, which is being tried out in real conditions the limits and characteristics of which are still not well known, at least quantitatively.

Strategies of choice to improve the robustness of fabric roofs include two approaches:

- Making the fabric stretchier so that the individual threads can solidarily resist a tensile force, instead of being ruptured one by one in a zipper-style fashion.
- Provide strong bands/lines/zones where a rupture will be arrested (see Section 6.6). Small defects will always occur, acting as *stress raisers* to propagate a zip. It is the goal of robustness to limit the extent of the rupture.

9.14 Falsework and scaffolding — a frequent type of failure

A well-known and notorious accident scenario, which keeps recurring stubbornly, involves the collapse of falsework under the weight of fresh concrete (see *Figure 9.15*). Its description is quite typical in most cases and can be summarized in a list of features, most of which have contributed to the failures on the past and keep doing so time and again:

- The work is performed in a competitive and profit-oriented environment, which results in time pressure and minimal supervision and quality control.
- On the same count, the reuse of worn, deformed and damaged material makes for rickety assemblies with load paths containing stress concentrations, unforeseen eccentricities, unequal and unsymmetrical loading, weakened materials and elements.
- Design notes, if any exist, usually cover the hypothetically ideal case of perfect alignment, considering weight as the only source of loading.
- Loads are transmitted through a series of elements placed on top of each other and held in position by contact pressure and friction only.
- The elements themselves (beams, vertical supports) are usually light and slender and bracings where provided are spaced far apart and attached by single bolt assemblies, often involving hinge-type details and looseness, which allow considerable movement before resistance sets in.

Fig. 9.15: Failure of a slab falsework during casting of the concrete

(a)

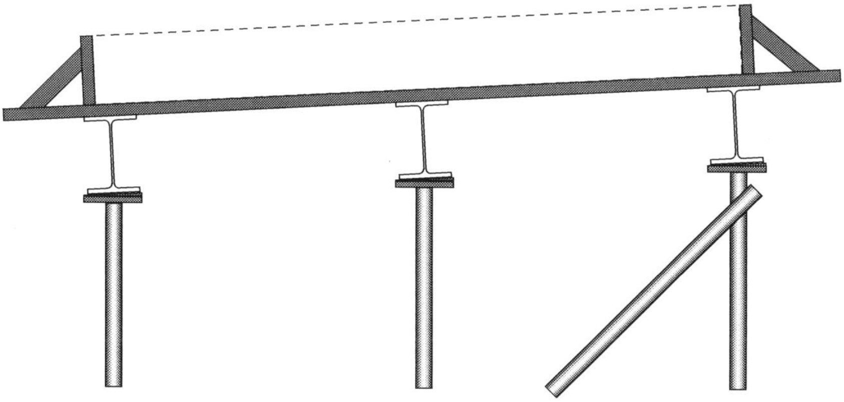

(b)

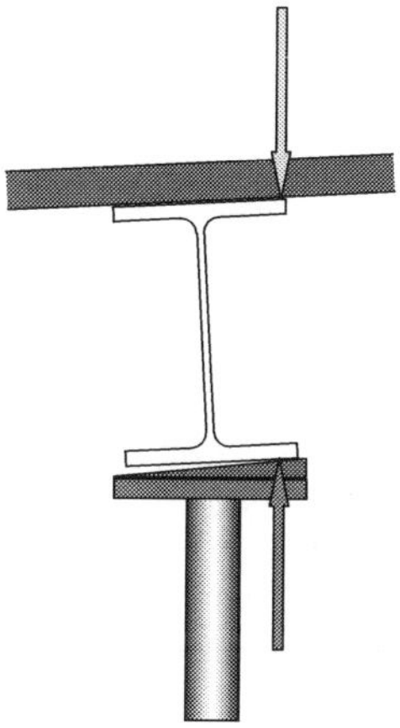

(c)

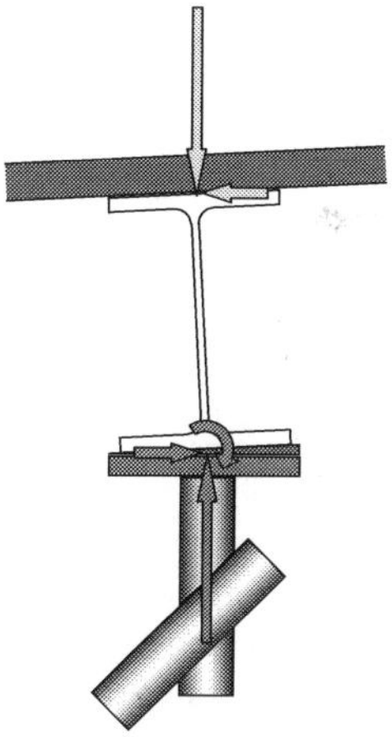

Fig. 9.16: Assembly and formwork for an inclined soffit; (a) general scheme. Transversal bending of beam due to, (b) inadequate or missing wedges, (c) transmission of horizontal forces

- The vertical supports are constituted by a multitude of slender posts, which will react in a brittle fashion when overloaded. The large number of elements and connection details makes the ensemble prone to omissions (missing bolts, pins or other deficiencies), each of which may trigger a progressive failure, given the brittle character of the elements.
- The fact that the structure is temporary – it may be effectively loaded during only about a day or so until the concrete has gained sufficient strength – makes for an attitude, where some tolerance towards safety margins seems to suggest itself. Building codes are not usually dealing with this sort of temporary structures.
- Loading situations during the placing and hardening of concrete involve force components other than vertical, due to gravity drift, lateral displacement of fresh concrete or movements of equipment. Often, the concrete being placed does not constitute a horizontal plate but the formwork is placed in a slope, with the supporting elements placed in that slope also, which for beams oriented at an angle to the slope will result in bending around the weak axis, as well as to the web itself in I-type beams (see *Figure 9.16*). Webs are usually quite thin and without stiffeners so that lateral buckling is one of the critical modes of instability.

- The imprecise adjustment of vertical supports, as well as the variable response of the soil when loaded, will result in unequal loading of some elements. Design notes, however, usually assume the load being shared uniformly by all load paths.
- Bracing is usually only provided to the vertical supports but not to the top part of the structure which consists of slender beams placed at more or less right angles in two layers. Sometimes they are loosely held in place by bent nails next to the bottom flange, or by clamps but often, no other elements provide resistance to lateral forces.
- Often, the assembly includes or consists of proprietary material, which is used repetitively many times. This had led to the illusion that something which has worked a number of times will also work this time, forgetting about seemingly minor differences in the many parameters and circumstances. "The jug goes to the well until it breaks" – this may not be so terrible involving a jug – translated to the failure of a scaffolding assembly, it may, however, involve loss of life.

Given this litany of ill circumstances, it is easy to understand how the persistent notoriety of falsework collapses has come about. Remedial action towards the various ills is not difficult to find, but it bears its cost, which will eat into somebody's profit.

Usually, an accident will serve to persuade the people and companies directly involved to take some action taking care of a number of deficiencies. However, a few years hence, or in a different location, companies, staff and circumstances will have changed and memory of the problems faded or disappeared.

The most serious single deficiency is usually the lack of lateral resistance and stability. Lateral forces may be minor compared with the weight of the freshly placed concrete, but they will always be present due to numerous causes as discussed above. We have learned as little children that there is a limit to how many building blocks and how imprecise and how loosely one can pile them one on top of the other, without the whole tower crashing down. Modern falsework assembly often resembles the play towers of long ago if not in scale then in systemic properties. Where the children's building blocks have rounded edges and slightly bulging faces, the steel, aluminium or wooden beams of the scaffolding assemblies have the same essential properties, due to wear and mishandling. Where precision of placing lacked due to the limited motor control in little children's hands, out of plumb and alignment is produced in falsework assemblies due to lack of attention and care, bent, deformed, worn, ill fitting or inconsistent elements, threads not working properly, etc.

These three approaches may provide most of the answers to the problem:

- Tight connections
  Provide connections, which allow no or only a minimum of movement/looseness. This will greatly reduce the ricketiness of the assembly and provide stability and resistance to secondary (i.e. lateral) forces.
- Reduce slenderness of elements
  For the vertical supports this may be done by decreasing the spacing of lateral bracing, or through a larger diameter of the posts. For the beams where lateral buckling or toppling is the critical mode of failure, cross elements could be introduced, similar to the lateral supports of timber joists or again, more sturdy members. Since more often than not falsework consists of proprietary elements, it will be onerous to provide additional features not provided in the supplier's kit. In this case, the only option may be to reduce the spacing of the supporting elements in order to supply more strength.

- Provide lateral bracing
  This is often done by attaching cables or diagonal struts to the top of the falsework, linking it directly to rigid elements such as concrete or steel work already built, or to the ground via auxiliary foundations or anchorages. Since the origin of the lateral loads is mostly in the surface of the formwork, it is at this level that lateral bracing should be attached. It should also and always be remembered that the lateral forces one tries to resist, can act in any direction, are unpredictable quantitatively by their nature of secondariness, and are also often of dynamic, shock-like character.

Web crippling has also been the main cause of falsework failures. Lateral movements accompany it, which makes its effects resemble the ones of bending in the weak axis and lateral buckling. Often it is found that the detailed description of events is a combination of the three.

For the design of a robust assembly therefore, careful consideration must be given to every element on the force path, and of course, to provide a force path for all possible loads.

As an example consider an assembly where the top layer of beams is placed at right angles to the slope of the soffit, of, for example 5%, i.e. 1/20 (see *Figure 9.16*). The beams will therefore be tilted at this angle, and a force component of 5% of the imposed weight will act across the top flange. To this an increment of force must be added, representing the effect of wind, activities taking place on the top, as well as gravity drift caused by dissymmetry in the load, or in the assembly itself. If all this lateral load must pass through the webs of the (usually very slender) beams, which may already be weakened by deformations and wear and tear, it is obvious that a critical situation exists which demands an additional load path in order to drain lateral loads away from the webs of the beams.

In the case of multiple-use of materials and elements, the question of conservation of robustness that was discussed in Chapter 1 becomes a key consideration. Construction being a rough place, every assembly and disassembly will be accompanied by some degradation, affecting certain pieces more than others. It is usually left to subordinate staff to pick out and eliminate material which is damaged, cracked, broken, worn, bent, torn, rotten, corroded or otherwise deficient, along with a tacit incentive to apply tolerances, as every piece thrown out represents a loss to the company. Inspection of the assembly is supposed to provide a check on this tendency but practically, it is often difficult or impossible to carry out thoroughly as there is no money, and more importantly, no time allotted for it. The inspector is typically expected to look at the works and approve it in maybe an afternoon, or a day – anybody who has walked through a forest of shores will know that it is dark in there, many of the hundreds or thousands of elements are partially or completely hidden from view. An experienced inspector will know when and where to look, but experienced inspectors are rare and far between, and they expect to be paid.

The important point here is not to lament these circumstances since they will hardly change in the future, but to find and use measures which mitigate or eliminate its consequences, in other words, to endow the system with sufficient robustness in the face of the adverse properties of the situation. These measures will likely be easier to inspect and correct than to detect flaws and defects in the maze of a typical scaffolding assembly.

In a recent example of a falsework failure, the forensic investigation found that almost all the features discussed above were present to various degrees, contributing to the collapse. No bracing or horizontal support to the top level of the assembly, i.e. the soffit form, had been installed which began to move horizontally when about one-third of the concrete for the massive slab had been placed and was being vibrated, triggering a sway movement of a large portion of the works.

Summarizing the essence of this typical case, it appears obvious that the problems of the intrinsic flaws of the structure, as well as of the presence of unpredictable lateral loads and forces are very prominent. The solution is found in the provision of additional force paths where resistance to those loads can be installed. Once the problem of lateral instability and resistance is taken care of, the multiple parallel beams and vertical supports can now attend to their duties, i.e. supporting the weight, without being impaired by force components they are ill fitted to resist.

9.15 Sabotage, how to mitigate its effects

Some recent terrorist attacks were targeted towards buildings, taking out substantial portions of the structure with the help of large amounts of explosives, causing widespread collapse. World politics being what they are, this will not likely stop in the foreseeable future and may reach any building, foremost the prominent and representative ones such as embassies, seats of power and finances, etc. where large and newsworthy loss of life and property can be achieved. In this chapter, only the explosives' attack on the ground floor shall be considered, since some other events such as the World Trade Center, NY ("9−11") have been exhaustively researched forensically and in terms of improvement of building design.

The intentional destruction of buildings and other structures, principally through knocking out of essential structural elements such as columns by means of explosives, will remain a preferred method so long as these elements are easily accessible, as well as easy to break with relatively small amounts of explosive or with clumsy techniques. This makes it cheap and achievable even for small groups of individuals without substantial financial backing, to cause great havoc and loss of life and property. It is well worthwhile to reflect on ways and methods to respond to this sorry fact of life.

A list can be made of types of defensive measures, some of which do not concern the structure itself at all such as the posting of guards, fences, obstacles, dogs, surveillance apparatus, alarm systems, etc. They work in certain circumstances and in others they do not. Adventure literature abounds of tales on how the hero with his cunning and a superhuman effort overcame the stratagem of the defenders and penetrated to the heart of his goal. Only, in our case, the hero is on the other side and is called a saboteur, terrorist or criminal.

Dogs and alarm systems being non-structural, let us assume that the scenario includes the hero/saboteur having penetrated to a proximity sufficiently close for him to execute his attack, be it with explosives, impact or similar means, perhaps not with the technical perfection he would be able to exercise given unlimited time but with sufficient effect to knock out one structural element such as a column or a piece of wall the size of a room. Typically, this will be on the ground floor and the element being attacked will be selected for maximum effect where again, this selection may not be based on the best of knowledge. The notion of decoys as a means to alleviate the threat comes to mind. By themselves decoys may not be very effective where inside knowledge is involved but combined with obstacles or sacrificial protective elements they may become a useful instrument of defence, deflecting the attack from the real target, which is one of the high ranking structural members – in this case most likely a column.

Without discussing the application of classic theory of fortification, the principle of structural survival in the face of more or less skilful terrorist attack can be summarized in the following manner:

Make access to the essential structural elements as difficult as possible through obstacles, disguises, decoys, sacrificial and protective barriers, enclosures or layers of expendable non-structural materials. The greater the distance between the centre of the explosion, and the hard surface of the structural member, the more the shock will be attenuated when it reaches its target. This is especially so when the space between is filled with materials whose destruction will require energy input: This energy will be deducted from the shock the structure itself receives, weakening it on its way to the target.

Use materials and structural detailing to provide toughness to the primary structural elements themselves in order to make destruction difficult. Typically, heavily reinforced concrete or composite steel/concrete members (e.g. steel jackets filled with concrete, or structural steel members cast into reinforced concrete) are difficult to knock out, requiring special placement and configuration of the explosive charges, which demands knowledge and skill and consumes time an attacker may typically not have. It was learned in the Second World War that such members could resist even large amounts of explosives if these are not placed precisely in the appropriate manner.

Provide duplication or multiplicity of load paths where possible. This may not always be easy to reconcile with architectural and functional requirements (transparency, openness, etc.) but if made an ingredient of the concept from the beginning, it may lead to acceptable features. For instance colonnades or clusters of several columns, leaving longer spans free between groups make it quite difficult to do effective sabotage as it takes many times the power needed to destroy one column, to knock out an entire group, where the individual members of a group are spaced, for example 1.5 m apart. Even if one or two out of a group were eliminated, it would be fairly inexpensive to provide the extra strength in the second floor structure so that the remaining elements could hold up the building – at reduced safety margins – until repairs can be performed.

Provide extra strength for the key elements. The provision of supplementary strength for the structural elements which are at once high up in the structural hierarchy, as well as exposed to attack (ground floor) conforms to a very old principle of building which wants that the key members (corner stone, key stone, etc.) be equipped with a higher standard of strength and quality so that the builder can rely on them with more certainty than on elements of lesser importance.

Reinforced concrete walls are more difficult to knock out than columns, provided the pressure created by an explosion is drained away elsewhere, through surfaces that are open or able to cede (knock out elements of the enclosure).

Chapter

10

Examples

10.1 Structural integrity of a historic building

This case concerns a complex of nineteenth century buildings located in a zone of moderate earthquake risk (Montreal, Canada). These buildings were constructed in the years 1860 to 1880 for the Sœurs de Miséricorde with the materials available at the time, namely stone for the walls and timber for the horizontal elements and columns. Most materials are reported to have been donated, which fact does not reflect well on the quality of the original construction.

In the course of their history, the buildings have gone through a number of major and minor modifications, mostly in connection with the installation of several generations of mechanical and electrical systems. Some of these modifications were less than thoughtful and left the structure in a state of severe degradation, with the telltales common to this style of construction: Walls bulging, beginning to buckle and disintegrate, floors sagging, floor joists broken and having to be shored up and portions of buildings evacuated and condemned because their condition had become too worrisome.

Robustness, or what part of it may once have existed, had obviously been lost. As well, structural integrity in its basic sense had never been provided, with the floor construction connected to the walls by contact friction only. In numerous places joists and floor beams were found to be sliding gradually off their supports, leaving the walls effectively freestanding for a height of two or several storeys (see *Figure 10.1*).

The buildings being of some heritage value, their demolition was deferred and studies undertaken to determine potential cost and circumstances of rehabilitation and new utility.

Rehabilitation will concentrate on re-establishing structural integrity, stability and robustness, as well as upgrading seismic resistance. Besides the repair of broken, damaged or degraded elements, it will include the installation of ties with some ductility anchoring the floor structure into the walls (see *Figure 10.2*) and some additional shear walls in the form of elevator and stair enclosures in reinforced concrete where the existing walls are found to be lacking sufficient strength to resist earthquake loads.

The floor finish can be used to create a diaphragm with some nominal reinforcing. The additional weight may have to be compensated through composite action or reduction of live loads. Surely, the structural integrity in terms of consistent and reliable coherence is the central concern in cases such as the one described above, which is typical for a very large quantity of older construction in most parts of the world, and the choice of strategy for robustness is fairly simple and straightforward: Buildings, where the connection of floors and walls is left to contact friction alone, will continue to disintegrate, the individual structural elements moving individually, slowly under the effects of gravity and quickly during an earthquake. Any rehabilitation must therefore include prominently the tying together of the various elements, vertical and horizontal.

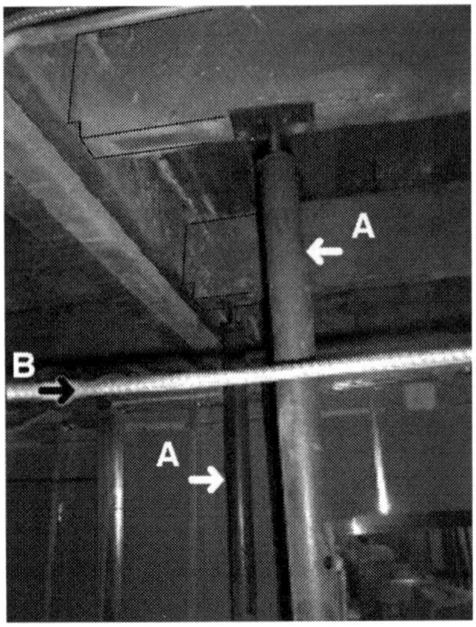

Fig. 10.1: Severely damaged floor. (a) Shores where support of joists is lost. (b) Tie rod to prevent wall movement

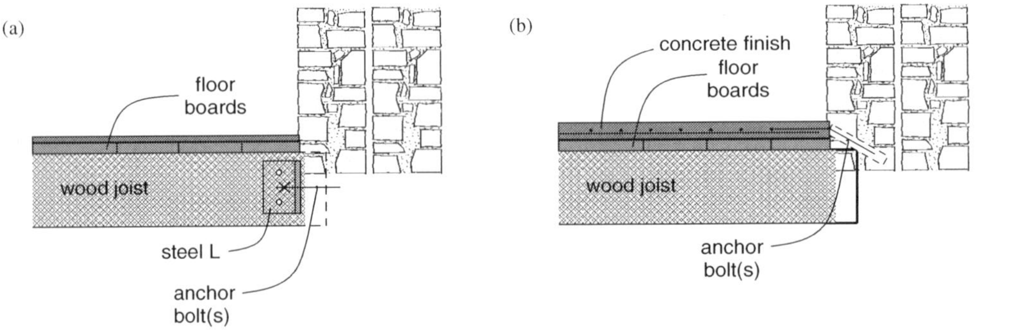

Fig. 10.2: Anchorage systems for floor structure and walls (a) with original floor level (b) with additional concrete finish

10.2 A cut-and-cover tunnel in a karst environment

The Adler Tunnel constitutes a more direct route than the old railway line that leads from Basel southbound to the central parts of Switzerland and through the Alps. The tunnel consists of a drilled core section of 4.3 km in length and cut-and-cover sections at each end. A general overview of the project with contributions covering design and construction is given in Ref. [5].

The western cut-and-cover sec tion which is 750 m in length traverses a gravel bed that was deposited by the Rhine River following the last glacial period. Beneath the gravel lies a Triassic, strongly karstic limestone stratum, known as Muschelkalk (shell limestone). This causes dolines (sink holes) and extensive subsidences to occur. Within the last 40 years about 20 new dolines formed with diameters between 5 and 10 m. According to estimations summarized in a geotechnical report dolines up to 22 m in diameter and 5 m in depth can form within a few hours and subsidences with a diameter of 100 m can develop with a settlement speed of 10 mm per year.

The Swiss Federal Railways as client and operating authority of the tunnel specified the following service requirements in the so-called utilization plan:

- A planned service life of at least 150 years.
- Partial replacements requiring lengthy closure of tracks not before the first 100 years.
- Speed limits of 160 km/h at present, 200 km/h in the future.
- Restrictive schedules for maintenance, taking into account time of day, number of tracks (single track or total closure) and intervals (daily, weekly, monthly, yearly closure).
- Acceptance of local moisture, no acceptance of dripping water that could form ice.
- Minimum acceptable radius of curved subsidence: 5000 m.
- Maximum subsidence: 250 mm.
- Maximum values of deflection for bridges spanning dolines.

Many of these requirements can be met by designing a robust structure.

Obviously, the slowly occurring subsidences can be taken into account by designing the cross-section with the necessary additional space which allows a realignment of the tracks (monitoring, correction). Suddenly occurring dolines, however, should be bridged by the tunnel acting as a stiff tube (strength).

In this case robustness not only means meeting requirements in the ultimate but also in the serviceability limit state (stiffness considerations).

In addition to the usual loads on a cut-and-cover tunnel the following scenarios were considered:

- A slowly developing extensive subsidence with an average speed of 10 mm per year and a maximum of 200 mm per year. These are caused by the collapse of caverns in greater depth.
- A superficial collapse of a karst cavern resulting in a crater reaching the surface or the tunnel level.

In both cases a reliable detection and subsequent refill are not possible within reasonable economical limits. The above cases were therefore assessed numerically in order to incorporate them as hazard scenarios in the structural analysis. *Figure 10.3* shows the two types of standardized subsidences (sinus shaped and with concentrated curvature) and the standardized spherical doline. The doline may occur centred or eccentric to the tunnel within hours. High precision in the design process is therefore not needed as the description of the scenario must rely on geotechnical knowledge and experience.

To bridge local dolines in any position, a monolithic tube over the whole endangered length was chosen. Any joints or hinges would impair the bridging effect.

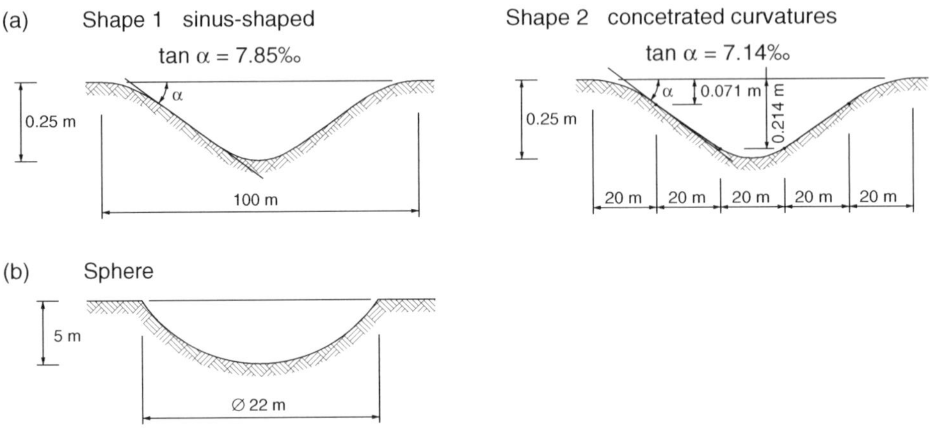

Fig. 10.3: Adler Tunnel; (a) standard subsidences shape 1 and 2, (b) standard doline [22]

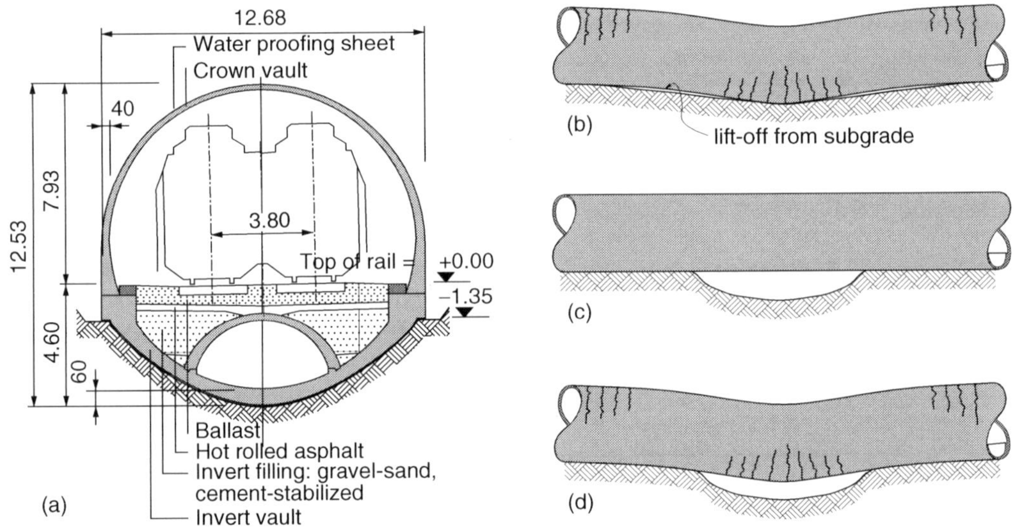

Fig. 10.4: Adler Tunnel; (a) typical cross section of the tunnel. Fundamental cases of structural performance: (b) extensive long-term subsidence, (c) doline, (d) subsidence with doline [22]

The structure chosen was an elastically supported tubular beam. For longer subsidences the tube will deform with the settlement. This situation can be treated as a serviceability problem; the deformations result in cracks and at most in the plastic hinges, but failure does not occur (see *Figure 10.4(b)*).

For dolines (smaller scale subsidences) the tunnel must bridge across and act as a tubular bridge (see *Figure 10.4(c)*).

It is very likely that the collapse of a deep cavern and the subsequent subsidence will lead to the failure of a cavern at a minor depth causing a doline to form simultaneously. The relevant and most plausible scenario to be considered is therefore the superposition of a typical subsidence and a doline (see *Figure 10.4(d)*). It is important to note that at the ultimate limit state the available deformability is reduced at the relevant cross section as it must already accommodate the subsidence. Therefore the behaviour of the tube spanning the doline is less ductile.

Structural analysis and design issues are treated in more detail in [22]. In the final design reinforcing bars Ø 26 mm with a spacing of 150 mm were used. Like this the ductile behaviour of the tunnel tube and its rotational capacity could be ascertained.

Although the projected hazard scenarios were supposed to occur only once in the service life of the tunnel, a similar incident happened already before the tunnel was commissioned:

In the middle of 1996 a sudden settlement of almost 200 mm was detected that had developed at a much higher speed than predicted. The intervals of control measurements were shortened to a fortnight and showed that the settlement proceeded with a speed of 1 mm/4 days. Cracks formed especially in the crown vault in a regular pattern with maximal widths of 0.2 mm.

Remedial action consisted of grout injections delivered from the ground above to a depth exceeding 60 m.

10.3 An avalanche gallery subject to train impact

The St. Gotthard auto route crossing the Alps in the centre of Switzerland follows the Schöllenen gorge between Göschenen and Andermatt. The same obstacle has to be mastered by the local narrow-gage railway, which operates with a toothed rack in steep sections. Normally the road and track run separately but one gallery serves for both lines protecting them from snow drifts and small avalanches (see *Figure 10.5*).

This gallery called *Nasse Kehle* was built from 1984 to 1986 as a composite structure consisting of a concrete back wall, steel columns and steel beams, covered with precast ribbed concrete slabs with a cast-in-place layer on top (see *Figure 10.6*). The spacing of both the columns and the transverse beams varies from 4 to 6 m depending on the assumed loading (see *Figure 10.7*).

The columns are exposed to possible impact from a derailed train or a run-away truck.

In the conceptual phase of design it was decided not to consider impact acting on the columns, but to install two additional rails within the normal track to prevent trains from derailing. In the detailing phase of the design, however, an additional longitudinal beam on top of the columns was provided to enable any column to be knocked out in case of an impact without overloading the rest of the structure. Consequently the columns were placed on hinges and fixed with small bolts only (see Section 6.9: *The Knock-Out Scenario*).

Due to planned flood protection measures for the adjacent Reuss River a structural examination was carried out in 2000 that produced the following results:

- The additional rails had never been installed because together with the toothed rack they would have hampered general maintenance of the track and especially snow ploughing in winter.
- Due to lack of space the guardrail between the sidewalk of the road and the railway line is suitable for light vehicles only.

Fig. 10.5: Avalanche gallery Nasse Kehle during execution

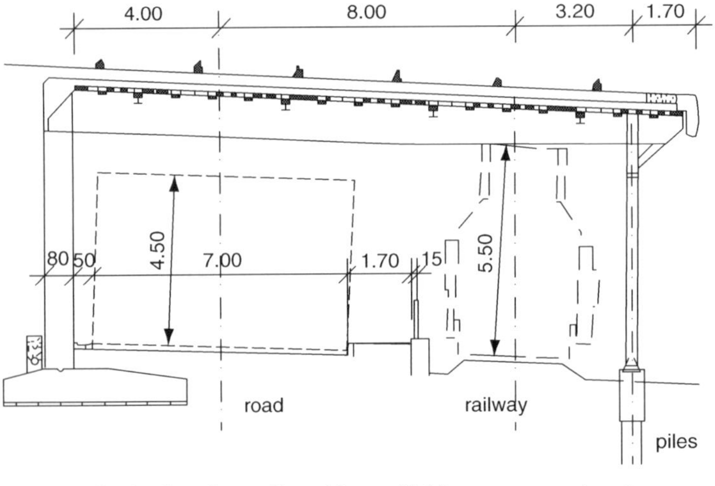

Fig. 10.6: Avalanche gallery Nasse Kehle; cross section (m, cm, according to [23])

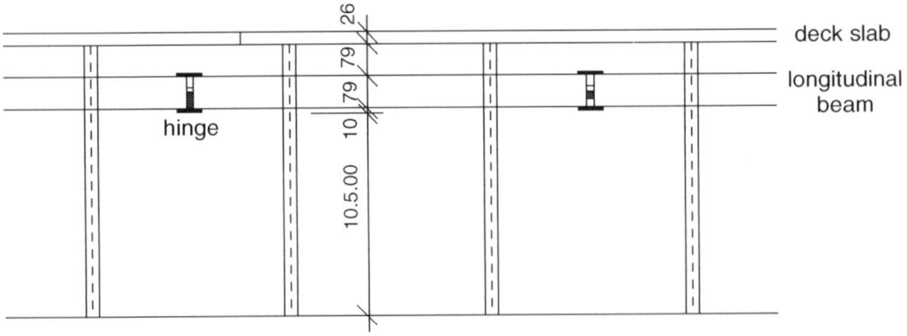

Fig. 10.7: Avalanche gallery Nasse Kehle; cut-out of longitudinal view (m, cm, according to [23])

- The largest settlements of footings amounted to 500 mm. Since some columns were not centred properly on the footings and others were out of verticality up to 2%, it must be concluded that settlements have begun already during construction and continued during operation of the gallery.
- In accordance with the general articulation concept that provided a joint every 20 m, the longitudinal beam on top of the columns was hinged, too. Since the hinges were placed at midspan, a knock-out of a column would have overstressed the longitudinal beam and led to an unacceptably large failure.
- Generally, the concept of allowing one column to fail was questioned, because even an engine that does not move forward anymore but only tilts would hit more than one column and at the most spoil them.
- The horizontal forces associated with an avalanche, combined with earth pressure on the retaining wall, were not compatible with overall stability criteria.

The remedial measures consisted of two phases, urgent safety measures undertaken prior to the next winter period and long-term measures executed one year later [23-26].

Here, only the evaluation of measures to cope with impact shall be discussed.

The list of possible constructional actions covered the following options (see *Figure 10.8*):

1. Additional rail outside the track to prevent derailing without hindering the snow plough (prevention).
2. Heavy safety barrier between road and rail (second line of defence, suitable for straying road vehicles only).
3. Longitudinal profiles fixed to the columns on different heights, tying them together, distributing longitudinal forces on several columns and acting as guidance to derailed engines or carriages (collision struts provide solidarization).
4. Improved load transfer between transverse beams and columns by additional lateral end plates to the columns (strength).
5. Encase the bottom part of the columns into a concrete wall improving both the flexural capacity of the upper parts of the columns and the resistance to direct impact on the lower parts (strength and solidarization).
6. Concrete wall along the track, placed on a slab beneath the track and connected with the footings of the columns (second line of defence).
7. Steel diagonals between the columns acting in compression in case of a knocked-out column (multiple load paths but reduced ductility for further unbalanced settlements).
8. Embankment or large rocks placed outside the concrete wall of option 6 and further improving lateral stability (strength).

Finally options 1, 3, 4 and 6 were combined to arrive at an economical and suitable solution that served for flood prevention as well. Like this it was possible to reduce probability and amount of impact forces to a minimum and to get a maximum of flexural capacity of the columns in the transverse direction resistance without reducing the over-all flexibility of the whole system in the longitudinal sense.

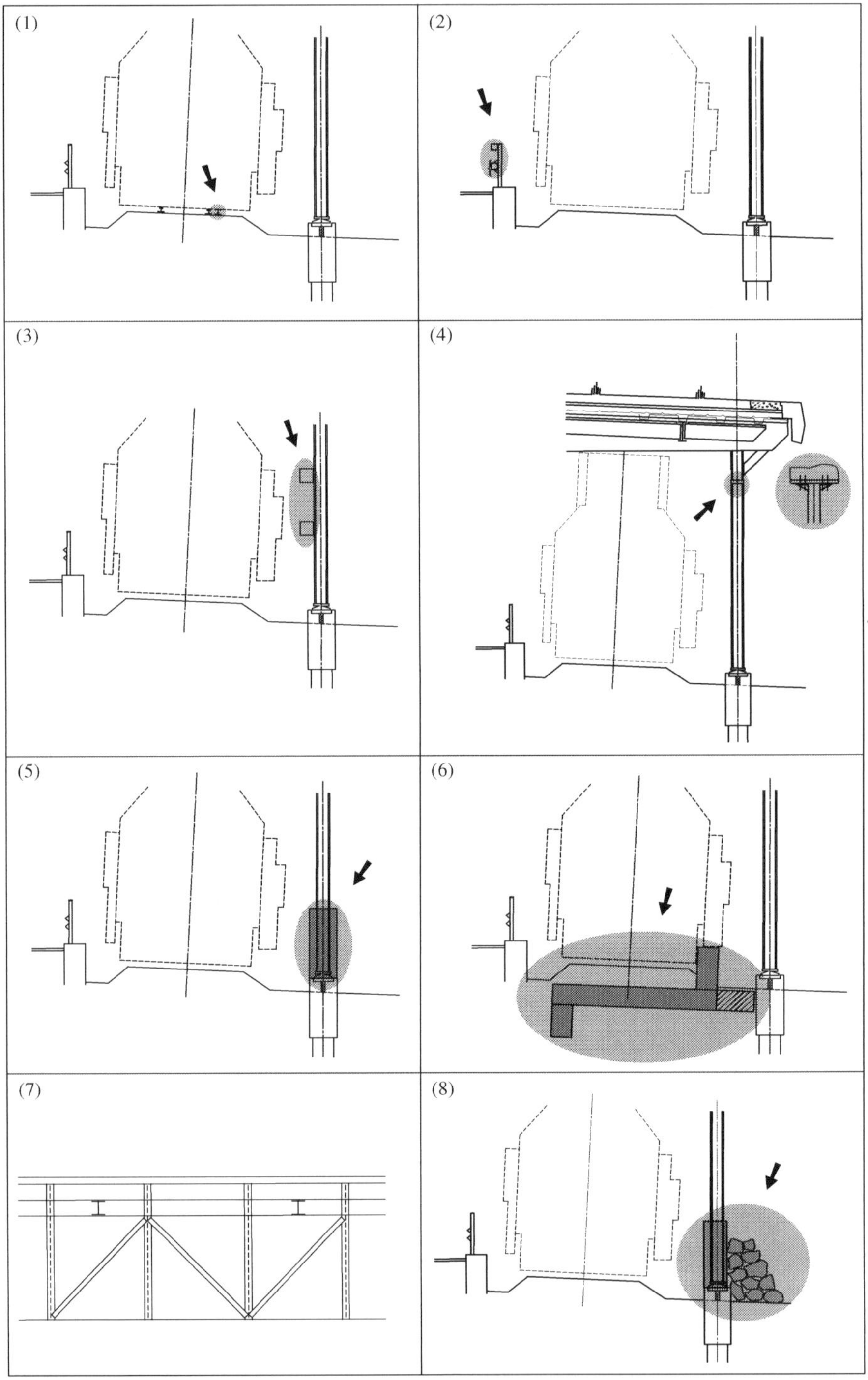

Fig. 10.8: Avalanche gallery Nasse Kehle; options to improve the safety of columns with respect to impact (according to [24])

10.4 Columns placed behind buffers (knock-out versus protective element)

The Zurich central railroad station was originally built as a terminal, track ends being equipped with buffers. During work done on the station building, the railroad needed temporary offices which were to be built in elevated position over the great concourse hall, on columns placed right behind the buffers (see *Figure 10.9*).

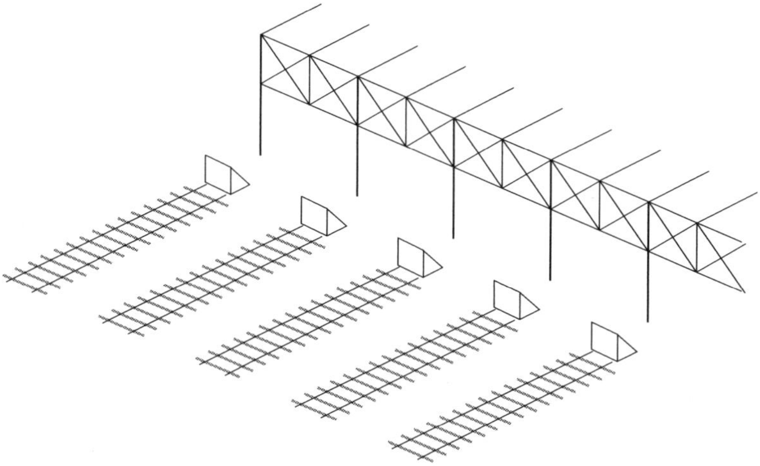

Fig. 10.9: Buffers or knock-out columns

The scenario of a locomotive over-running the end of the track, pushing the buffer away, was considered in the design concept and two options were recognized:

- To make the buffer strong enough.
- To prepare the columns as knock-out elements, with the structure above providing a second line of defence by bridging over two spans.

Practical considerations (logistics, cost, etc.) then led to the choice of the second option, which resulted in the addition of diagonals in the second storey façade of the building. The columns were attached at both ends with bolted assemblies designed to rupture at a low shearing force.

The option of a strong buffer resisting the impact would have been an equally straightforward solution to the robustness problem through providing a protective or sacrificial structure. However, the existing buffers were not up to the task and would have had to be rebuilt much stronger, involving massive foundation work and disruption of the operation of the station.

10.5 The case of a cable-stayed bridge

The bridge indicated in *Figure 10.10* was built for two-lane road traffic. Each stay was made from four 3.5 inches (90 mm) diameter rods, arranged in a block of four, with threaded ends and a standard nut and washer seated on a simple plate assembly.

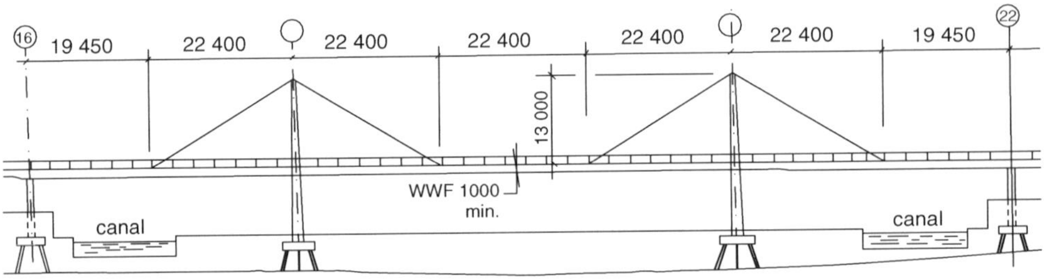

Fig. 10.10: Pont des Îles, Montréal, Canada

In a severe snow storm ($T = -20^\circ$C) during night, three single rods in three different locations ruptured at the point of attachment. The diagnosis found that the failure was the compound effect of a number of contributing causes, the most important ones being:

- Inferior quality steel with a transition temperature above the ambient temperature experienced during the storm making the steel brittle.
- Badly cut threads.
- Seating details restraining rotation of the rods, causing secondary bending stresses.
- Extremely low damping.
- Wake buffeting of the leeward rods (which were moving in and out of the wind shadow of the windward members (*Figure 10.11*)).

The mechanism of failure was determined to be low cycle fatigue (between 20 000 and 50 000 cycles) due to wind-excited vibration.

The fact that four units had been provided for the stays in each location prevented the collapse of the bridge which continued to carry light road traffic for about two hours after the damage had been first noticed.

The event in this case was the combined effects of a particular phenomenon of aerodynamics and of several flaws in the creation of the bridge.

The circumstances of the failure could/should have been anticipated and corrected or eliminated, which did not take place for lack of special knowledge and experience. However, robustness had been provided in the form of multiple (four) parallel members (i.e. load paths) in each location, so that the loss of one out of four units did not cause any further damage or collapse.

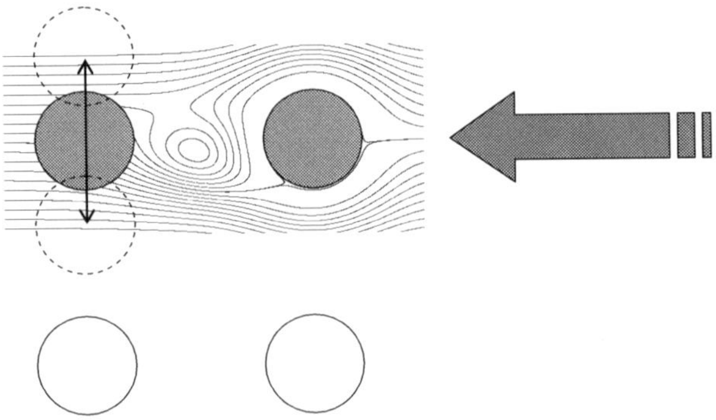

Fig. 10.11: Cross wind vibrations due to wind

Subsequent to the event, all stays were replaced with better quality steel, the seating detail was improved to permit rotation and the four individual rods of each stay were tied together with a clamp assembly, which effectively prevents vibration.

Modern stay cable bridges tend to be made much more robust having a fan- or harp-shaped arrangement of a number of stays along the span.

10.6 Power lines

In early January 1998, a weather condition called *freezing rain* brought down most of the electric transmission lines in the south-west part of the province of Quebec; adjacent parts of the province of Ontario and of the states of Vermont and New-York were also affected. This weather condition is not rare at all as it occurs every winter, sometimes more than once, affecting some portion of this northern land. It was unusual only in two respects that two *freezing rain* events followed each other in close succession, the intensity of the cumulative effect exceeding the ordinary, and that a more populous area was struck: About 3 Mio people were left without electricity during one week to more than one month, in the middle of the Canadian winter, until repairs – some of them makeshift style – were completed.

The immediate cause for the collapse of so many power lines (more than 10 000 towers and masts fell to the ground) was the ice which accumulated on the wires and cables so massively that examples of more or less cylindrical bodies of solid ice up to 6 inches (150 mm) in diameter were discovered, which represents an additional weight of about 11 lbs/foot (16 kg/m) on each wire. This weight would then have to be supported by first the conductors and cables in catenary fashion, and by the masts in the form of vertical, and in the case of unequal load or spans, or of ruptured conductors, also horizontal loads near the top of the mast.

More modest amounts of ice had been anticipated and included in the rules and procedures for the design of the conduits, wires and towers, but extremes as experienced in 1998 had not generally been considered, for obvious reasons: Numerous power lines must be built in inaccessible parts of the land and weight comes therefore at a high premium, moreover the mere quantity of the infrastructure demanded extreme economy and optimization.

What is of interest in the context of robustness is not that failures do occur at loading levels, which it would be uneconomical to consider. However, two aspects of this major disaster (according to [14] at least 25 deaths indirectly related to the event, many from hypothermia), economic losses of many billion $ and a considerable amount of discomfort merit to be discussed:

- The lattice-type towers (see *Figure 10.12*), which were used for most of the major transmission lines are structural systems of very low forgiveness, in other words with little ductility and therefore prone to fail in brittle fashion. Due to geometric requirements relating to the economic and geographic constraints noted above, towers must be as distant from each other as possible, which in turn makes them very tall – the catenary of the conductors must be at a safe height above ground. They must also be as light as possible. All of this had made the space truss the preferred answer to the task, with very slender members made from high-strength steel (slenderness $\lambda = l/i$ up to 150 or even 200), which tend to fail suddenly and completely by elastic buckling, i.e. in a brittle fashion. The space trusses of the towers having very little redundancy, brittle failure of any member will be paramount to brittle failure of the entire

Fig. 10.12: Domino-style failure of power lines

structure. This was in fact observed as some motion pictures of collapsing masts were taken.

- The second, more important observation concerns the amplitude of the failure, which included power lines of over 100-km length which failed in domino-like fashion, tower by tower being pulled down by the weight of the ice, and, even in the absence of excess weight, by the one-sided horizontal loading caused by the collapse of the tower next to it. This was the main reason why it took a very long time to reconstitute the operation of the system, even though work crews from a large part of North America were called in to help. Domino-like progressive failures are the most important type of failure that the design for robustness must prevent. Following the catastrophic failure of the power supply in the region around Montreal (Quebec), a commission was appointed with the task of identifying the causes and reasons for the event, its amplitude and circumstances such as the organization and implementation of disaster relief, etc., as well as to find ways and means to prevent or mitigate similar events in the future, all of this within reasonable limits of expenditure.

Concerning the first observation, the excessive slenderness (i.e. brittleness) of the structural elements, it was noted that masts with a unique shaft made from a tapering steel tube apparently offered much better resistance to the extreme loading conditions created by the ice storm. From a review of the structural situation of such elements, it is quite clear why this is so: No buckling modes with high degrees of equivalent slenderness exist for a tubular structure while for a lattice-type space truss, every member (mostly lattice towers are composed of single angles, for ease of assembly) represents a (mostly elastic) buckling mode, with some additional modes involving larger portions of the structure.

The domino-style failure was made the subject of intensive discussions, the conclusion of which was, unequivocally, that *strong points* must be provided at intervals of five to ten spans where the progressive (i.e. domino) effect is intercepted and stopped. This must be done in the form of *robust* towers, which will resist all loads, even in extreme conditions (i.e. in any case), including excessive weight and the one-sided pull caused by the failure of one of the neighbours, while the other one is still standing. The relatively low additional expenditure for the stronger (i.e. heavier) construction of one tower in five or ten is henceforth considered a good investment in the context of the North American electricity system.

The case of the power line towers provides one of the most clear-cut and classically simple examples of what robustness is about and how to achieve it to an acceptable degree:

To prevent failures in extreme conditions may in some instances be too costly. Limiting the spread of failure then becomes the goal of design for structures.

References

Literature referred to in the text

[1] Damgaard Larsen, O. (1993). Ship Collision with Bridges—The Interaction between Vessel Traffic and Bridge Structures; *Structural Engineering Documents* No. 4, April 1993, IABSE Zurich, 132 pp.

[2] Dodd, G. (2004). Structural Glass Walls, Floors and Roofs; *Structural Engineering International*, 14(2), pp. 88–91.

[3] EN 1990 (2002). Eurocode—Basis of Structural Design; CEN, 2002, 89 pp.

[4] EN 1991-7 (2006). Eurocode 1: Actions on structures: Part 1–7: Accidental Actions; CEN, 2006, 66 pp.

[5] Grether, M. (editor, 1996). Adlertunnel; *Schweizer Ingenieur und Architekt*, Nr. 18/1996, pp. 337–370.

[6] Gulvanessian, H.; Vrouwenvelder, T. (2006). Robustness and the Eurocodes; *Structural Engineering International*, 16(2), pp. 167–171.

[7] Haldimann, M.; Luible, A.; Overend, M. (2008). Structural Use of Glass; *Structural Engineering Documents*, No. 10, January 2008, IABSE Zurich, 215 pp.

[8] Hess, R. (2004). Material Glass; *Structural Engineering International*, 14(2), pp. 76–79.

[9] Kott, A.; Vogel, T. (2004). Safety of Laminated Glass Structures after Initial Failure; *Structural Engineering International*, 14(2), pp. 134–138.

[10] Knoll, F. (1982). Human Error in the Building Process; *IABSE Journal*, J 17/82.

[11] Knoll, F. (1984). Modelling Gross Errors; Risk, Structural Engineering and Human Error; Waterloo Press, Waterloo, Canada.

[12] Knoll, F. (1986). Checking Techniques; Modeling Human Error. Edited by A. Novak., ASCE, Ann Arbor Michigan.

[13] Knoll, F. *et al*. (1983). Summary; Rigi Workshop, IABSE, Switzerland.

[14] Meteorological Service of Canada (MSC). The Worst Ice Storm in Canadian History; Homepage of MSC,
http://www.msc-smc.ec.gc.ca/media/icestorm98/icestorm98_the_worst_e.cfm

[15] NFPA 98 (2002). Guide for Venting of deflagration; US National Foundation of Prevention of Accidents.

[16] Peyer, B. (1985). Deckeneinsturz im Hallenbad Uster vom 9. Mai 1985—Orientierung der Bezirksanwaltschaft Uster vom 29. Mai 1985 (Ceiling collapse in the Uster swimming hall of May,9, 1985—Briefing of the district attorney of May 29, 1985); *Schweizer Ingenieur und Architekt*, Nr. 23/85, pp. 566–568.

[17] Rice, P.; Dutton, H. (1995). Structural Glass; E & FN Spon, London, 144 pp.

[18] Schittich, Ch. *et al*. (2007). Glass Construction Manual; 2nd revised and expanded edn., Birkhäuser Basel, 352 pp.

[19] Schober, H.; Schneider, J. (2004). Developments in Structural Glass and Glass Structures; *Structural Engineering International*, 14(2), pp. 84–87.

[20] Schuler, C.; Bucak, Ö.; Albrecht, G.; Sackmann, V.; Gräf, H. (2004). Time and Temperature Dependent Mechanical Behaviour and Durability of Laminated Safety Glass; *Structural Engineering International*, 14(2), pp. 80–83.

[21] The Institution of Structural Engineers (1999). Structural Use of Glass in Buildings; London, 168 pp.

[22] Vogel, T.; Kalak, J. (1998). Cut-and-cover Tunnel in a Karst Environment; *Proceedings*, IABSE Colloquium Stockholm 1998 'Tunnel Structures', IABSE Zurich, pp. 289–294.

[23] Wolf, Kropf & Bachmann AG; Plüss Meyer Partner AG (2000a). Lawinengalerie Nasse Kehle; Bauliche Sofortmassnahmen, Massnahmenprojekt (Avalanche Gallery 'Nasse Kehle', urgent safety measures, preliminary design); Swiss National Roadways, Highway Department Canton Uri, 6.10.2000, unpublished.

[24] Wolf, Kropf & Bachmann AG; Plüss Meyer Partner AG (2000b). Lawinengalerie Nasse Kehle; Massnahmenkatalog Thema Aussenstützen (Avalanche Gallery 'Nasse Kehle', options for remedial measures, topic columns); Swiss National Roadways, Highway Department Canton Uri, 22.11.2000, unpublished.

[25] Wolf, Kropf & Bachmann AG; Plüss Meyer Partner AG (2001a). Lawinengalerie Nasse Kehle; Dossier zum Massnahmenkonzept (Avalanche Gallery 'Nasse Kehle', dossier of conceptual design); Swiss National Roadways, Highway Department Canton Uri, 09.02.2001, unpublished.

[26] Wolf, Kropf & Bachmann AG; Plüss Meyer Partner AG (2001b). Lawinengalerie Nasse Kehle; Dossier zum Massnahmenprojekt (Avalanche Gallery 'Nasse Kehle', dossier of preliminary design); Swiss National Roadways, Highway Department Canton Uri, 02.03.2001, unpublished.

Other literature

Agarwal, J.; Blockley, D.I.; Woodman, N.J. (2001). Vulnerability of 3D Trusses; *Structural Safety*, 23(3), pp. 203–220.

Agarwal, J.; England, J.; Blockley, D. (2006). Vulnerability Analysis of Structures; *Structural Engineering International*, 16(2), pp. 124–128.

Alexander, S. (2004). New Approach to Disproportionate Collapse; *The Structural Engineer*, 82(23), pp. 14–18.

Bailey, C.G.; Toh, W.S.; Chan, B.M. (2008). Simplified and Advanced Analysis of Membrane Action of Concrete Slabs; *ACI Structural Journal*, 105(1), pp. 30–40.

Baker, J.W.; Schubert, M.; Faber, M.H. (2008). On the Assessment of Robustness; *Structural Safety*, 30(3), pp. 253–267.

Barmish, R. (1994). New Tools for Robustness of Linear Systems, MacMillan Publishing, New York.

Beeby, A.W. (1999). Safety of Structures and a New Approach to Robustness; *The Structural Engineer*, 77(4), pp. 16–21.

Boulding, K.E. (1989). Towards a Theory of Vulnerability; *Journal of Applied Systems Analysis*, 16, pp. 11–17.

Burnett, E.F.P. (1975). The Avoidance of Progressive Collapse: Regulatory Approaches to the Problem; National Bureau of Standards, Washington, D.C.

Callaway, D.S.; Newman, M.E.J.; Strogatz, S.H.; Watts, D.J. (2000). Network Robustness and Fragility: Percolation on Random Graphs; *Physical Review Letters*, 85, pp. 5468–5471.

Department of Defence (2005). Unified Facilities Criteria (UFC): Design of Buildings to Resist Progressive Collapse; Department of Defence (DoD), Washington, DC, 2005.

Ellingwood, B.R.; Leyendecker, E.V. (1978). Approaches for Design against Progressive Collapse; *Journal of Structural Division*, 104(3), pp. 413–423.

Ellingwood, B.R.; Dusenberry D.O. (2005). Abnormal Loads and Progressive Collapse; *Computer-aided Civil and Infrastructure Engineering*, 20(5), pp. 194–205.

Ellingwood, B.R. (2006). Mitigating Risk from Abnormal Loads and Progressive Collapse; *Journal of Performance of Constructed Facilities*, ASCE 20(11), pp. 315–323.

Ellingwood, B.R. (2007). Strategies for Mitigating risk to Buildings from Abnormal Load Events; *International Journal of Risk Assessment and Mitigation*, 7(6/7), pp. 828–845.

General Services Administration (2003). Progressive Collapse Analysis and Design Guidelines for New Federal Office Buildings and Major Modernization Projects; General Services Administration (GSA), Washington, DC, June 2003.

Grierson, D.E.; Xu, L.; Liu, Y. (2005). Progressive-Failure Analysis of Buildings Subjected to Abnormal Failure; *Journal of Computer-Aided Civil & Infrastructure Engineering*, 20, 155–171.

Haddon, W. (1980). The Basic Strategies for Reducing Damage from Hazards of all Kinds; *Hazard Prevention*, 16, pp. 8–12.

IStructE (2002). Safety in Tall Buildings; Institution of Structural Engineers, London.

Izzuddin, B.A.; Tao, X.Y.; Elghzouli, A.Y. (2004). Realistic Modelling of Composite and Reinforced Concrete Floor Slabs Under Extreme Loading I: Analytical Method; *Journal of Structural Engineering*, ASCE 130(12), pp. 1972–1984.

Izzuddin, B.A. (2005). A Simplified Model for Axially Restrained Beams Subject to Extreme Loading; *International Journal of Steel Structures*, 5, pp. 421–429.

Izzuddin, B.A.; Vlassis, A.G.; Elghazouli, A.Y. *et al*. (2008). Progressive Collapse of Multi-storey Buildings due to Sudden Column Loss—Part I: Simplified Assessment Framework; *Engineering Structures*, 30, pp. 1308–1318.

Lamont, S.; Lane, B.; Jowsey, A.; Torerro, J.; Flint, G. (2006). Innovative Structural Engineering for Tall Buildings in Fire; *Structural Engineering International*, 16(2), pp. 142–147.

Leyendecker, E.R.; Ellingwood, B.R. (1977). Design Methods for Reducing the Risk of Progressive Collapse in Buildings. *Building Science Series*, 98, National Bureau of Standards, Washington, DC.

Lind, N.C. (1995). A Measure of Vulnerability and Damage Tolerance; *Reliability Engineering and System Safety*, 48, pp. 1–6.

Lind, N.C. (1996). Vulnerability of Damage-Accumulating Systems; *Reliability Engineering & System Safety*, 53(2), pp. 217–219.

Lu, Z.; Yu, Y.; Woodman, N. J.; Blockley, D.I. (1999). Theory of Structural Vulnerability; *The Structural Engineer*, 77(18), pp. 17–24.

Maes, M.A.; Fritzsons, K.E.; Glowienka, S. (2006). Structural Robustness in the Light of Risk and Consequence Analysis; *Structural Engineering International*, 16(2), pp. 101–107.

Müllers, I.; Vogel, T. (2008). Dimensioning of Flat Slab Structures for Column Failure; *Structural Engineering International*, 18(1), pp. 73–78.

National Institute of Standards and Technology (2005). Final Report of the National Construction Safety Team on the Collapses of the World Trade Center Towers; NIST NCSTAR 1, Draft for Public Comment, National Institute of Standards and Technology, USA, September 2005.

Office of the Deputy Prime Minister (2000). The Building Regulations 2000, Part A, Schedule 1: A3, Disproportionate Collapse, 1992 Edition, Fourth Impression (with amendments) 1994, further amended 2000, Office of the Deputy Prime Minister, London, UK.

Radowitz, B.; Schubert, M.; Faber, M.H. (2008). Robustness of externally and internally posttensioned bridges; *Beton- und Stahlbetonbau*, 103(S1), pp. 16–22.

Rausand, M.; Hojland, A. (2004). System Reliability Theory (second edition); John Wiley & Sons, New York.

Smith, J.W. (2006). Structural Robustness Analysis and the Fast Fracture Analogy; *Structural Engineering International*,16(2), pp. 118–123.

Song, L.; Izzuddin, B.A.; Elnashai, A.S.; Dowling, P.J. (2000). An Integrated Adaptive Environment for Fire and Explosion Analysis of Steel Frames—Part I: Analytical Models; *Journal of Constructional Steel Research*, 53, pp. 63–85.

Sorensen, J.D.; Christensen, H.H. (2006). Danish Requirements for Robustness of Structures: Background and Implementation; *Structural Engineering International*, 16(2), pp. 172–177.

Starossek, U. (1999). Progressive Collapse Study of a Multi-Span Bridge; *Structural Engineering International*, 9(2), pp. 121–125.

Starossek, U. (2006). Progressive Collapse of Structures: Nomenclature and Procedures; *Structural Engineering International*,16(2), pp. 113–117.

Starossek, U. (2007). Typology of Progressive Collapse; *Engineering Structures*, 29(9), pp. 2302–2307.

Starossek, U. (2008). Avoiding Disproportionate Collapse of Tall Buildings; *Structural Engineering International*, 18(3), pp. 238–246.

Taylor, D.A. (1975). Progressive Collapse, *Canadian Journal of Civil Engineering*, 2(4), pp. 517–529.

Val, D.V.; Val, E.G. (2006). Robustness of Frame Structures; *Structural Engineering International*, 16(2), pp. 108–112.

Vlassis, A.G.; Izzuddin, B.A.; Elghazouli, A.Y.; Nethercot D.A. (2006). Design Oriented Approach for Progressive Collapse Assessment of Steel Framed Buildings; *Structural Engineering International*, 9(2), pp. 129–136.

Wada, A.; Ohi, K.; Suzuki, H.; Kohno M.; Sakumoto, Y. (2006). A Study on the Collapse Control Design Method for High-Rise Steel Buildings; *Structural Engineering International*, 16(2), pp. 137–141.

Wu, X.; Blockley, D.I.; Woodman, N.J. (1993). Vulnerability of Structural Systems—Part 1: Rings and Clusters; *Journal of Civil Engineering System*, 10, pp. 301–317.

You may order SED monographs, or get information on other IABSE publications, directly from:

IABSE, ETH Zürich, CH-8093 Zürich, Switzerland
Phone: +41-1-633 2647 Fax: +41-1-633 1241
e-mail: secretariat@iabse.org web: www.iabse.org
Order Form: www.iabse.org/publications/orderform

Structural Engineering Documents 10
Structural Use of Glass, by Matthias Haldimann, Andreas Luible, Mauro Overend

Published: May 2008; 215 pages; ISBN 3-85748-119-2

Structural Engineering Documents 9
Cable Vibrations in Cable-Stayed Bridges, by Elsa de Sa Caetano

Published: October 2007; 188 pages; ISBN 978-3-85748-115-4

Structural Engineering Documents 8
Use and Application of High - Performance Steels for Steel Structures,
by Hans-Peter Günther

Published: October 2005; 152 pages; ISBN 3-85748-113-7

Structural Engineering Documents 7
Use of Fibre Reinforced Polymers in Bridge Construction, by Thomas Keller

Published: August 2003; 131 pages; ISBN 3-85748-108-0

Structural Engineering Documents 6
Structural Bearings and Expansion Joints for Bridges, by Günter Ramberger

Published: May 2002; 89 pages; ISBN 3-85748-105-6

Structural Engineering Documents 5 - Revised Version
Introduction to Safety and Reliability of Structures, by Jörg Schneider

SED 5 revised version (2006) includes Variables Processor software, supporting the subjects treated ***free:*** educational version VaP 1.6;
one-year free trial: professional version VaP 2.2

Published: second edition 2006; 109 pages; ISBN 3-85748-093-6

About the Authors:

Dr. Franz Knoll is Vice-President of Nicolet Chartrand Knoll, structural engineers, in Montreal, Canada.
He received his doctoral degree (1966) and an honorary doctorate (2000) from the Swiss Federal Institute of Technology (ETH) in Zurich in connection with work on structural safety, and as a structural designer.

Thomas Vogel is Professor of Structural Engineering at the Swiss Federal Institute of Technology (ETH) in Zurich since 1992. He got his diploma in 1980 at ETH and has worked as a consultant for more than ten years.
Beside robustness, his research interests cover the conservation of existing struc- tures, non-destructive testing methods and ductile design with brittle materials.

Structural Engineering Documents

Objective:

To provide in-depth information to practicing stuctural engineers in reports of high scientific and technical standards on a wide range of structural engineering topics.

Topics:

Structural analysis and design, dynamic analysis, construction materials and methods, project management, structural monitoring, safety assessment, maintenance and repair, and computer applications.

Readership:

Practicing structural engineers, teachers, researchers and students at a university level, as well as representatives of owners, operators and builders.

Publisher:

The International Association for Bridge and Structural Engineering (IABSE) was founded as a non-profit scientific association in 1929. Today it has more than 3900 members in over 90 countries. IABSE's mission is to promote the exchange of knowledge and to advance the practice of structural engineering worldwide. IABSE organizes conferences and publishes the quarterly journal Structural Engineering International, as well as conference reports and other monographs, including the SED series. IABSE also presents annual awards for achievements in structural engineering.

For further Information:

IABSE-AIPC-IVBH
ETH Zürich
CH-8093 Zürich, Switzerland
Phone: Int. + 41-44-633 2647
Fax: Int. + 41-44-633 1241
E-mail: secretariat@iabse.org
Web: www.iabse.org